PRIVATE INTERESTS, PUBLIC POLICY, AND AMERICAN AGRICULTURE

STUDIES IN
GOVERNMENT AND PUBLIC POLICY

PRIVATE INTERESTS, PUBLIC POLICY, AND AMERICAN AGRICULTURE

WILLIAM P. BROWNE

UNIVERSITY PRESS OF KANSAS

Published by the University Press of Kansas (Lawrence, Kansas 66045), which was organized by the Kansas Board of Regents and is operated and funded by Emporia State University, Fort Hays State University, Kansas State University, Pittsburg State University, the
University of Kansas, and Wichita State University

Library of Congress Cataloging-in-Publication Data
Browne, William Paul, 1945–
    Private interests, public policy, and American agriculture.
    (Studies in government and public policy)
    Bibliography: p.
    Includes index.
1. Agriculture and state—United States.
2. Agricultural industries—United States.
3. Food industry and trade—United States.
4. Lobbying—United States. 5. Pressure groups—
United States.   I. Title.   II. Series.
HD1761.B73   1988   338.1'873   87-23131
ISBN 0-7006-0334-4
ISBN 0-7006-0335-2 (pbk.)

British Library Cataloguing in Publication Data is available.

Printed in the United States of America
10 9 8 7 6 5 4 3 2 1

To

all those residents of the Sioux Valley who taught me at the outset that the realities of agricultural policy are part of life. Their faces change, as do the farm conditions and public policy circumstances they confront. And to John Schlebecker, who, while lecturing on agricultural history, first began to help me put those three facts together.

Contents

Preface

There are more reasons than I, at any one time, can recall for writing about the representation of agricultural interests. Three sets of these reasons are so essential to this book that they must be acknowledged at the outset, however. The first—and perhaps the most obvious—of these reasons relates to contemporary political and economic events. Politically, the past ten years have seen a general renewal of farm activism at both the grassroots level and in Washington, D.C. Agrarian protest once again has come to political center stage as farmers have forcefully articulated the story of their financial hard times. The term "crisis" has been repeatedly used in describing the economic conditions that trouble farm and farm-related industrial producers. Someone, therefore, needed to examine how agricultural interests have responded as their concerns expanded beyond the long-standing questions of what to plant, buy, and borrow to budget deficits, the high value of U.S. currency, international trade prospects, high interest rates, and declining land values. There are, as David Truman taught us, at least some linkages among the conditions in which people find themselves, the public policy responses they see as appropriate to their conditions, and their political behavior. Because agricultural production has become so reliant on government programs, the time is right to judge the relationships among these three variables with respect to agriculture.

Agricultural interests also should be studied because they represent such an important component of the modern American economy. In terms of sheer numbers of workers, the agricultural sector as a whole employed 18.5 percent of the available civilian work force in 1984. The products and commodities that these workers create and market are extremely diversified. Agriculture, for both production and distribution purposes, has come to rely on capital-intensive and high-technology inputs rather than on its previous small farm/small business labor-intensive operation. Because of its increasing productive capacity, which has heightened chronic problems of product oversupply, American agriculture functions in a competitive world market rather than just a domestic one. These conditions have created opportunities for the emergence of more and more specialized spokespersons in the agricultural public policy arena. A great deal needs to be understood about the complexities of where, when, and how such a diverse array of individual activists, organized groups, corporate lobbyists, and consultants articulate the many problems and policy needs of agriculture and agribusiness and of how the agricultural sector in particular and American business in general operate interdependently with government.

A second set of reasons for studying agricultural interests has as much to do with the past as the present. Neither agricultural policy nor agricultural interests are new. The linkages among farm problems, public policy responses, and group activism are long-standing ones. Moreover, scholars have frequently used agriculture-based explanations in generalizing about American politics, especially in the area of interest groups. One of that small cadre of mutually supportive political scientists who now, as we put it, "do agriculture," best presented this historical rationale. At one of those long-forgotten academic conferences, Ken Meier noted just how much of our discipline's collective knowledge of interest groups was gained from the study of farm interests. There are the specific agrarian works of such scholars as Grant McConnell and Theodore Saloutos that remain required reading on the evolution of interest representation in the United States. In addition, such important contemporary group theorists as Theodore Lowi, Mancur Olson, Robert Salisbury, David Truman, Jack Walker, Graham Wilson, and Harmon Zeigler focused quite substantially on agriculture as they developed their generalized explanations of interest group behavior. Much of their analysis is now dated, descriptive of only a few interests, or very limited in the degree to which it brings together agricultural policy decisions and the forces that attempt to shape them. So this book needed to be written and read to update our view of a rapidly changing sector that has long received important academic attention.

In expanding on previous scholarship, this book has one unifying theme that places agricultural interests in historical perspective. Through my research, I have come to believe that the programs and benefits of agricultural and food policy are so extensive and so specifically targeted to particular types of users that no single interest can be concerned with all of them. Interest representation is fragmented along policy lines within agriculture and among the various decentralized legislative and administrative policymakers responsible for different programs. Lobbyists and farm activists act only on those issues that directly affect what they see as their programs. As a result, they often do business quite differently, with little or no awareness of what concerns other agricultural representatives.

The same behavior goes on over time. In response to changing conditions, interests generate new activity. New lobbies emerge. Older ones have new or different policymakers on whom to concentrate. Still others redefine their demands on government as well as the strategies they find useful in attempting to attain their political goals. There has long existed, I believe, an ongoing evolution of interest identity and behavior in any specific policy area. In agriculture, protest and social groups were joined by general farm organizations, which later found themselves facing commodity organizations, trade associations, and then corporate representatives as patterns of production, distribution, and consumption shifted.

Groups and organizations of all types have persisted because the public policies of interest to them often have been only slightly changed rather than redefined every few years. Organized interests, along with other participants in the policy process, work to maintain themselves, their political relationships, and their constituency support by protecting previous policy gains as well as by creating demands for new and modified programs. Thus, this book has been written as a reminder that representative interests in an economic or social sector are not just orderly and businesslike forces that address only the specifically agreed-upon needs of the here-and-now. Nor are they simply anachronistic vehicles for re-articulating what have become the myths of another era. Such interests must be a little of both. They are defenders of the status quo that nonetheless must also work for new, more relevant programs.

My final reasons for studying this topic are personal. After studying interest groups for my entire professional career, I remain fascinated by agricultural representation less because of its lessons about American politics than because of the circumstances of my birth and socialization. My awareness of politics and public policy is due to childhood lessons taught by Iowa farmers and rural businesspeople whose successes and failures depended on the farm economy. As political scientists have long noted, agrarian environments have bred the country's most active and often most volatile electors. All of my early years were spent with people who alternately praised and condemned farm politics. These people believed that politicians affected the quality of their lives. I learned from them that public policy decisions influenced financial ledgers, new purchases, the migration of friends to urban areas, and a host of small things that even a child could see in changing fields, feedlots, stores, and homes. While others of my generation may attribute their political education to the civil rights movement, Vietnam, and Watergate, I look back instead at the Soil Bank, commodity prices, and my neighbors' prevailing views of Ezra Taft Benson.

My intense enthusiasm for the topic does not mean that the research was conducted to reaffirm my identity, extend blame, or praise rural America. I remain by nature very much a political observer and very little a political activist. My personal interests have simply kept me, as a political scientist, very attuned to agricultural politics, so in a time of agricultural change and protest, it just seemed to make sense that I be the one to undertake this project.

THE STUDY AND ITS DATA

This book results from research that took place in four stages. During stage one, in early 1977, I undertook an evaluation of four major general farm organizations in response to commentaries on their loss of influence. Twelve formal staff interviews concerned group services, membership expectations, and the impact of the membership on organization lobbying.[1] Stage two was an attempt to analyze the extensiveness of the farm and agribusiness lobby. Congressional hearings and written records were examined, and fifty open-ended interviews with lobbyists and officers were conducted. I sought to find out who these organizations were, what problems they faced, and what they saw as their basic interests.[2] The 1977 and 1981 farm bills provided a framework for this research, which was conducted in 1977, 1978, and 1981. During this time, a new farm lobby appeared that used protest tactics uncharacteristic of contemporary farm lobbies. The American Agriculture Movement (AAM), unlike the groups that I had been studying, did not want to work with government. I spent stage three during parts of 1978 and 1979 following the AAM around the country, asking 161 leaders what they really wanted from government and how they planned to get it.[3] As in the earlier stages of research, a great deal of additional time was spent "soaking and poking" not only with group activists, but also with a variety of congressional and USDA personnel. Stage four began with three years of assembling information and trying to figure out what I had learned. I kept asking questions of agricultural policymakers and lobbyists to help sort things out. By 1985, I felt ready to conduct more extensive interviews. Information from this survey constitutes the primary data base for this book.

THE SAMPLE

The questions posed in the preceding section were used in selecting a sample of organizations and respondents as well as preparing a questionnaire. The sample was purposely selected from, first of all, groups and firms that I had seen, heard, and read about as especially active in agricultural policy. This provided an initial list of 128 organizations with widely varying concerns about all three dimensions—farm, food, and trade—of agricultural policy. The list included general farm organizations, commodity groups, farm corporations, agricultural protest groups, rural organizations, agribusiness firms, trade associations, general business groups and firms, unions, consumer organizations, environmental and other public interest groups, foundations and research institutes, and consulting firms. In June 1985, by numerical lot, I randomly selected a sample of 40 percent (51) of the organizations from my alphabetized list for initial interviews. A second sample of 40 percent was also selected in case additional interviews seemed useful and time allowed for them.

I also planned, at the time, to interview a sample of twenty-five less active agricultural organizations from my own list of "other lobbies." Twenty-two of these organizations were selected in October and November on the basis of referrals from legislators, legislative staffers, and Department of Agriculture employees.

I also made plans for additional interviews with representatives of other active lobbies from my main list that were involved in activities that were newsworthy, of unusually great concern to other respondents, or otherwise unique. On this basis, eight organizations were selected from September through October. Representatives of 130 organizations were interviewed, 122 in person and 8 on the telephone. (See Appendix A.)

In addition to collecting information about individual groups and firms, I wanted to examine the behavior and policy involvement of several coalitions of interests. Nine well-established coalitions were selected for detailed study on the basis of visibility, longevity, political prominence, and variety.

THE INTERVIEWS

Respondents were chosen on the basis of their direct responsibilities for representing the organization. Usually the principal lobbyist, executive officer, or consultant was selected. In larger organizations, executive directors and those actually responsible for agricultural legislation were interviewed.

In almost half of the organizations more than one respondent was selected. Additional contacts were chosen if the initial interviewee suggested it, if more detailed information was necessary, or if new questions came up during the interview process. Because of their representatives' vitriolic opinions about existing farm policy, protest groups were treated somewhat differently. To look for evidence of consistency and continuity in the policy preferences of grassroots activists, interviews were held both before and after the passage of the Food Security Act of 1985. A total of 238 respondents participated in the interviews, 52 of them from protest groups. No one turned down requests for interviews.[4]

The interviews followed a series of open-ended questions prepared in May. Most interviews took place in the respondent's office, home, or farm. Interviews lasted from fifteen minutes to three hours. The average length of an interview was one hour and fifteen minutes.

The first interviews were held with grassroots activists and state organization officials in June and July 1985. Washington representatives were interviewed from September through early January 1986. Some return interviews were conducted the following March. Final interviews with protest activists in the Midwest were conducted in January 1986.

THE QUESTIONNAIRE

The questionnaire used for the interviews was tightly structured in order to assemble an aggregate data base representing responses from each organization (see Appendix B for interview outline). Many probe questions were included to allow flexibility in pursuing important points of interest about policy involvement. The format also made for relaxed interviews during which the respondents felt free to elaborate on their own and others' successes and failures.

The interviews began by asking respondents to define their organization's interest in agricultural policies and issues of the 1980s. What did they set as priorities? Which bills, amendments, and administrative rulings did they deal with? Respondents were then asked to describe their lobbying strategies and the degree to which they were active on each of these priority issues.

Two problems were anticipated. First, there would be few written records. Lobbyists often reacted on a moment's notice when informed of an issue or policy problem of likely concern and, afterwards, promptly moved on to something else. Second, the question, Who represents what? would not always be easy to answer. In the early stages of research, I was surprised at the number of times that a group was inactive on a policy question that seemed a matter of obvious interest. Why, given a general interest in the topic at hand, were their representatives not actively lobbying on the proposal? This question led me to gather data on issue representation and nonrepresentation.

To control for these problems, I randomly sampled 180 USDA and congressional decisions made between 1980 and 1985.[5] These were classified into nine categories: price supports/farm income, trade and markets including food aid, taxation, agricultural structure, subsidized inputs, conservation/food security, nutrition/domestic hunger, rural America/poverty, and health and safety (see Appendix C). After respondents answered questions on priorities, I asked about their involvement with specific types of issues that they had not mentioned. Questions varied for each organization, depending on the type of interest initially identified. If, for example, a firm's primary interest was in export markets, I asked about two to three unmentioned decisions from that specific category. When interests spanned two or more categories, I asked two or three questions about each interest. I asked about the raising, mobilization, and adequacy of lobbying resources; relationships with policymakers from other organizations, Congress, and the administration; coalition activity; the political strength of various types of interests; and the events and participants in the 1985 farm bill debates.

In many instances, often at the beginning of our conversation, respondents volunteered much, if not most, of this information. This meant that many interviews deviated from the sequence of questions. It also contributed to the need for multiple interviews for many of the organizations and follow-ups with some respondents. In a few instances, some questions were left unanswered.

Acknowledgments

This work exists only because of the support and assistance of a great variety of people. The most important are those activists and lobbyists who spend their lives shaping American agricultural policy. Members of Congress, their staffs, and an assortment of United States Department of Agriculture (USDA) officials were no less helpful. The concern and attention of a list of individuals so long that a directory would be needed to include them all is the foundation for this study. They were my near-constant mentors from 1976 through 1986. Only my excursions into other research projects, the classroom, and Lake Michigan provided them relief from my presence. These people have not only my thanks but my apologies as well for the time and energy they expended in answering my questions.

More specific thanks to a few others who can more easily by noted. Policy participants provided the essential substance for this analysis, but there was other critical assistance. The Economic Research Service (ERS) of the USDA, the Farm Foundation, and Resources for the Future funded related projects that were responsible for my learning about domestic agricultural and food policy, the relationships of trade and foreign policy to agriculture, agricultural research and technology issues, and rural policy. The ERS was always at the center of these efforts, and I must credit the ERS staff for providing me an education of the highest quality.

The Agricultural and Rural Economics Division and the National Economics Division of the ERS funded time in Washington to conduct interviews during the fall congressional debates on the 1985 Farm Bill. The ERS also provided clerical support from the Farm and Rural Economy Branch, an office in the Rural Business and Government Branch, and some time for data analysis and writing. John Lee, Ken Clayton, Ken Deavers, Bob Bohall, David Brown, Norm Reid, Dave

Harrington, and Tom and Barbara Stucker made all this happen and opened other staff doors in furthering my education in the content of agricultural programs. Wayne Rasmussen did the same from a rare and detailed historical perspective that I could not have done without.

Resources for the Future, through its National Center for Food and Agricultural Policy and the support of the W. K. Kellogg Foundation, provided a grant to buy that time necessary to finish my writing in the fall of 1986. Beyond this financial support, former Center Director Kenneth J. Farrell offered continued advice on my research. He made a marked difference in my ability to explain what I had observed.

My own institution, Central Michigan University, provided a year's leave through a sabbatical and a University Research Professor Award. There was also a summer fellowship and travel support for conference papers on the project. People there did far more than provide me with funds, however. In particular, I owe great thanks to Delbert J. Ringquist, Myron Henry, and Douglas Friedrich, who used their administrative positions to remind me that scholarship matters, that it will be rewarded, and that it can be done in an environment where it is often forgotten. Also at CMU, I have a great debt, which I will never even attempt to repay, to Joann Gust and Rosemary Thelen, who managed the clerical aspects of my grants and manuscript. Douglas Spathelf, who manages CMU's grants and contracts, was similarly helpful.

Of course, no manuscript could be completed without the thoughtful criticisms of academic colleagues. Several were especially helpful in reading drafts and near-final copy of this book and related articles. The comments of James T. Bonnen were the most useful, since he read and commented on the entire manuscript. Jeffrey M. Berry, Allan J. Cigler, Don Hadwiger, Larry Hamm, Michael Hayes, James Guth, Harold Guither, Christopher Leman, Kenneth J. Meier, Richard Miller, John Peters, Henry J. Pratt, Robert H. Salisbury, Ross B. Talbot, and Charles W. Wiggins reviewed parts of the manuscript and were always helpful. Salisbury and Wiggins can never be thanked enough for the years they spent in watchful critiques of my work. Mark Lundgren and John Dinse contributed to the book's content. ERS colleagues also commented in helpful ways on both my study design and final projects. Finally, the staff of the University Press of Kansas did all the editorial things necessary to get the final product out.

I take personal responsibility for any mistakes in judgment or lack of adequate explanations. These numerous friends only brought me assistance; neither they nor anyone else could ensure my infallibility in reporting and analyzing what I observed. In answering any criticisms of this work, all I can say in advance is: That's the way I saw it. With that thought always in mind as I wrote this book, I continuously sought solace in the immortal words of Jimmy Buffet, "Don't try to describe the ocean if you haven't seen it, you just might end up being wrong." There is no better explanation as to why I kept up the soaking and poking when others suggested I take it easy.

1

Introduction

This book, by design, is unlike others written about interest groups. Most academic writings on interest groups fall into three categories, all of which I aspire to avoid in order to further our understanding of interest group behavior. First, there are broad theoretical treatments of interest representation. David B. Truman's *Governmental Process* and James Q. Wilson's *Political Organizations* best represent this research genre.[1] Such studies have done much to show what it means for private interests to be an integral part of the political process. In their explanations, interest group theorists draw heavily on examples to illustrate how both organizations and policymaking work.

The tendency of the major group theorists to support conclusions through selective examples has proved a mixed blessing. On the positive side, it has produced a surprisingly broad range of conceptual works covering most aspects of group behavior. Important studies range from behavioral analyses of how members are attracted and held loyal to the group to macro explanations about how interests exercise power. Theorists have left few gaps in developing linkages between the organization of interests and their influence on policymaking.

The negative side of this sort of group theory relates to the degree to which linkages have not been satisfactorily established. Are the specific events cited just too self-serving as examples of normal group behavior? Do things usually happen that way? Those are the questions that critics justifiably ask even as they acknowledge the importance of conceptual understanding.

Interest group theory, as a consequence, spawned two distinctive categories of research as analysts sought both to elaborate and to test basic assumptions of interest group theory. The second and third types of published works on interest groups are either case studies of single organizations or empirical behavioral studies of some universe of individuals involved with interest organizations. Both types of microlevel research have been important in supporting interest group theory and in adding to it. Unfortunately, critics have been all too successful in challenging both approaches and often call into question on methodological grounds the importance of previous findings.

The case study literature is by far the most encompassing of these research types. When researchers have avoided the time-bound "how a bill becomes a law" format, they have produced richly detailed accounts of what takes place. Andrew S. McFarland's case study of Common Cause, for instance, goes far beyond examples in explaining the routines, tactics, strategies, and sources of influence of that particular lobby.[2] Because of his immersion in the life of the organization as he conducted his research, McFarland was able repeatedly to demonstrate why specific behavioral patterns developed and why they sometimes change. But is Common Cause typical of Washington lobbies? The inability of case studies to offer comparative observations and data causes the harshest critics simply to dismiss them. Even the most sympathetic proponents of the case approach acknowledge the limited degree to which case study finds can be generalized to other organizations and policy settings.

Empirical studies with large data bases do offer generalizability as to patterns of behavior among group members, organization officials and lobbyists, and even policymakers who are targets of influence. Lester W. Milbrath's often-emulated *The Washington Lobbyists* remains, nearly twenty-five years after its publication, the classic behavioral study of interest groups.[3] Milbrath skillfully used data to build and support his theory of lobbying as a communications process. As a result of his work, political scientists have come to accept that lobbying demands access to policymakers and the transfer of information to them within the overall process that Truman had laid out earlier. Empirical researchers since Milbrath have added contextual dimensions to their studies to test for behavior differences among policy types and among states.[4] Researchers have also reexamined behavior based on the different types of interests represented.[5] However, even these studies have not silenced the many critics who continue to point out how few systematic data support group theory.[6]

By pointing to weaknesses in data, critics are rather paradoxically addressing the greatest strength of the even-more-maligned case study. Because of the great wealth of detail in analyses of single organizations, it is often possible to discover what lobbying accomplishes for the participants and those they represent. When the collection of independent variables must be limited and set forth in uniformly applicable questions for empirical research, such explanations and discoveries are rarely possible. So, although empirical research on interest groups has the advantage of greater generalizability and the collection of information about entire populations of policy participants, its disadvantage has been the relative superficiality of its conclusions about what lobbying accomplishes. Nonetheless, both macrolevel theorizing and microlevel data collection have contributed a great deal to a collective understanding of interest groups and lobbying, and the credibility of this work should not be denied.

Research by Henry J. Pratt has suggested a way of minimizing the problems of the two research types by combining their methodological approaches.[7] Pratt studied each of the nationally organized Washington lobbies in a single issue area, aging policy. He gathered empirical data on political participation and compared it for his small universe of groups in order to provide a more comprehensive micro foundation for previous macro theories. The major problem with the design of Pratt's study was the number and kind of organizations represented. There were but a handful of groups, and they did not represent the variety of lobbies, both those staffed by professionals and those maintained by volunteers, usually described in the literature.

The research for the project described in this book proposed a far broader inquiry into one of the primary sectors of the U.S. economy, one that involved interests from the largest possible number of social groups. Agriculture, as one of the oldest organized sets of interests involved in influencing U.S. public policy as well as possibly the single largest sector of the economy,[8] is arguably the best possible choice for such a study.

WHY AGRICULTURE AS A POLICY ARENA?

There was a time when agricultural policy meant farm policy. The policy process attended to three problems of maintenance: maintaining farm incomes, maintaining a farm capacity to support needed food and fiber supplies, and maintaining a rural resource base of land and soil. That time has passed.

Since long before the 1980s, agricultural policy has dealt with a far more complicated array of issues than those of basic farm maintenance. Issues of nutrition, safety, quality, and domestic assistance have become institutionalized in agriculturally related legislation. Food Stamps and other food aid provisions are always a major concern in the periodic renewals of omnibus farm bills. The twelve food aid programs accounted for nearly $18.9 billion in expenditures for fiscal 1985, a total of approximately 26.2 percent of total United States Department of Agriculture (USDA) appropriations.[9]

The magnitude of food, as contrasted with farm, programs can be seen by looking at aggregate retail food costs. Although about 35 percent of food prices are accounted for by costs of production and farm profits, only 12 percent of which is generated on the farm, the remainder is associated with the various distribution and marketing costs of bringing commodities from the farm to the consumer. This entire middleman process involves inspection, regulation, direct recipient support, and taxation, all of which compound the relationship between farm and government.

Farm and food program concerns have more than just a domestic component. Since 1973, U.S.-grown commodities have been exported at an annual rate of 33 percent of the value of farm production with peak sales of about 38 percent. Foreign sales began to fall in the early 1980s as competitors in other countries increased their production capacity and ability to compete, while U.S. exports were simultaneously troubled by high prices associated with the internationally strong U.S. dollar. The same economic conditions made imports more attractive to the domestic food market.

Export and import considerations now dominate the foreign policy dimensions of agriculture. Economic policies affecting currency rates, technical assistance to foreign governments, the removal of U.S. barriers to foreign trade, and import restrictions have come to be part of the puzzle of finding a market for American farmers and those who handle their products.

International food assistance for emergency needs and for developmental assistance has become yet another component of the maze. In the Food for Peace Program, USDA spent $2.1 billion in fiscal 1985. Most of the funds went to support purchases and distribution of U.S.-produced food and farm supplies. To accommodate foreign assistance and trade efforts, USDA operates several agencies exclusively for international exchange. Cooperative efforts also involve the State Department and the Agency for International Development. Such food trade and assistance programs have important implications for rewarding U.S. allies, threatening punishment of its foes, and, more generally, affecting the internal economic conditions of both recipient countries and other international traders. As a consequence, foreign policy aspects of food trade often take on a life of their own, well beyond the export-oriented agenda of those concerned with international aspects of American agriculture.

Agriculture policy has three specific dimensions of farm, food, and foreign issues.[10] As a policy arena, it is large enough and so diversely structured that it serves and affects numerous groups. Although fewer than 3 percent of U.S. residents work on farms, nearly one in five workers is involved in some facet of farm production and supply or food and fiber distribution and service. Workers involved in these enterprises contributed 17.9 percent of the gross national product for 1984.[11] Agricultural exports have been so important that, as recently as 1980, they created a surplus of $23.9 billion in reducing a total trade deficit of $48.7 billion from the nonagricultural sector. Even in 1985, after food exports had declined 23.5 percent since 1984, agricultural products still accounted for 15 percent of the financial value of total U.S. exports.[12]

Defined this way, the agricultural lobby represents most economic and social interests. Agricultural policy is still most visibly addressed by producer organizations. However, agribusinesses involved with the delivery and export of farm products are also well organized. There are firms that supply farmers and others that buy and distribute farm products. Many do both. These firms lobby on their own, but even more frequently through specialized associations representing such interests as fertilizers, agricultural chemicals, grain traders, commodity refiners, processors, grocery manufacturers, and retailers. Nearly all agribusinesses involved with food, fiber, and tobacco products are represented in national politics.

Other business interests besides those in the production and distribution chain find agriculture policy of great importance. Lending and investment institutions, for example, have invested heavily in agriculture as well as in foreign countries that can be affected by trade policy. Many U.S. firms with foreign investments hope to stabilize political events in those countries through agricultural development. For these same economic reasons, foreign-based commodity organizations sometimes seek representation in U.S. agricultural policy, and foreign governments may attempt to voice their opinions directly.

In addition to manufacturing and commerce, agribusiness is also represented through union employees. The United Auto Workers union maintains a specific farm equipment division. Other national unions, such as the Teamsters and the American Federation of Labor-Congress of Industrial Organizations (AFL-CIO), have organized local branches with predominantly agricultural emphases. Farm workers are organized as well. In response, so are farm employers.

Consumers are organized around many food issues. There are different organizations that center on nutrition, the poor, rural residents, overseas relief, and school children. Social causes such as world peace, conservation, the environment, and human rights are organized around agricultural policy. Patrons of these reform groups include churches, private foundations, and individual members from every segment of society.

All of these interests play some role in the three dimensions of agricultural policy. Some are more active than others. Some have very little opportunity to exert any influence. These organizations do, however, constitute a collective agricultural lobby that is extremely diverse, representative of most economic and social interests, and capable of adding great complexity to the policy process.

Another reason that makes this an excellent research setting for studying private interests is that the agricultural policy arena is in a condition of flux. Historically, agricultural policymaking focused on the agricultural committees of Congress and the United States Department of Agriculture. Agricultural interests concentrated all their attention there. In the mid-1980s, this is only a partially accurate description. Although major farm legislation remains influenced primarily by traditional agricultural policymakers in Congress, the USDA, and private interests groups, others, which have become integral parts of the process, cannot be discounted. (See Table 1.1.) On commodity price support legislation for the 1985 farm bill, a variety of institutions vied for influence. The Office of Management and Budget worked for budgetary cuts, for example. The Office of Trade Relations served as a foremost proponent of unrestricted international trade and lower U.S. prices. Private interests, as only one set of policy partici-

TABLE 1.1
THE 1985 FARM BILL PARTICIPANTS:
COMMODITY PRICE SUPPORTS

Congressional agriculture committees	Conservation interests
Farm State Congressional Delegations	Congressional appropriations committees
General farm organizations	Office of Management and Budget
Commodity organizations	Office of Trade Relations
Commodity user coalitions	Commodity divisions within the USDA
Individual project consultants	Office of the Secretary, United States
Foundation/agribusiness study reports on	Department of Agriculture
agricultural conditions	White House

pants seeking to affect legislative outcomes, had to contend with *all* of them.

On issues such as food safety and taxes, agricultural interests have entered an even more expanded policy process with many more institutional participants. Most of these new participants—at least new to agriculturalists—have had little experience with farm policy. The result is a series of legislative and administrative contests in such unfamiliar places as the House Ways and Means Committee or the Food and Drug Administration. In these settings, traditional agricultural interests not only contend for influence with nonagricultural interests but also often with hostile administrative experts and such incompatible policy conditions as escalating budget deficits as well.

In summary, agriculture is an excellent choice of policy arenas because of its economic importance to the United States as an integrated production sector, the wide range of interests represented in agricultural policy debates, and the transitional status of agricultural policymaking. On the one hand, agriculture represents an issue area in which an old style of close and established relationships dominates policymaking. On the other hand, agriculture is a vast policy arena in which contestants are unfamiliar with one another or with the policymaking locale. Therefore it also represents a newer and more open style of policymaking in which friendship patterns seldom prevail. Lobbying within this issue area probably confronts the broadest array of political problems possible, and its practitioners probably employ the widest gamut of strategies and tactics found anywhere.

PURPOSE OF THE BOOK

This book was written for three distinct audiences: those concerned with agricultural policy, readers who simply want to know more about interest groups, and, of course, political scientists. In attempting to satisfy each set of readers, there may well be more information here than any one individual will find useful. It is no less likely that many of the intended readers will find less written here than they want. Those in agriculture might wish for more policy analysis. Scholars of politics may criticize this book for the failure to touch upon all relevant dimensions of group theory. The only defense for omissions of either sort is that one cannot include everything in a relatively short, readable book that should be useful to more than a highly selective audience. Both to avoid disappointment and to guide the reader through the volume, its essential concerns must be delineated. Seven questions directed the research that provided material for the book.

How is the Agricultural Lobby Organized?

From a policymaking perspective, it has become very difficult to tell who the players are and what issues and interests they actually represent. Many policymakers believe that the number of organizations that represent agricultural issues is growing annually, especially with the renewal of each new farm bill. New groups are emerging. Established organizations not previously active in agriculture are now involved. In addition, multiple representation of single interests has become more common. Organizations not only employ more lobbyists on their own staffs; they also hire more lobbyist consultants from private firms. Several groups also bring in officials and employees of state affiliates during especially busy times. It should be clear that this book is about more than interest groups as they are commonly defined. The focus is not merely on voluntary associations but on any private, nonparty organizations. This expansion of diversely organized private interests is having a serious effect on public policymaking because so many nontraditional interests are being expressed.

But how accurate are general perceptions of change? Are these distinguishable trends or merely isolated examples of expansion in the agricultural lobby? Recent research in national politics suggests that growth patterns are very real for the Washington lobby as a whole.[13] The number of registered lobbyists, including part-timers and non-Washingtonians, more than doubled between 1976 and 1986—from 3,420 to 8,800. Approximately 20 to 25 percent of these lobbyists represented some type of farm, food, fiber, or related trade issue. There is no information on the relationships between the policy activists who represent private interests, however.

This research void leaves several gaps in our knowledge about lobbying. To what extent does representational expansion simply reflect the development of a more complex lobby rather than the actual articulation of more distinct points of view? It may be that many organizational attempts are little more than ways of adding to the resource base of established groups and firms. Agribusinesses may open Washington offices to enhance the lobbying efforts of those with whom they share a common interest. Some new organizations may serve as little more than fronts for interest groups that wish to become politically active in agricultural policy without the attendant visibility. How important are consultants in advancing strategies, gaining access, and just generally improving lobbying efforts? Do representatives from the states bring similar contributions to Washington lobbyists?

WHO REPRESENTS WHAT IN AMERICAN AGRICULTURAL POLICY?

There is no research that provides extensive data linking the representation of private interests to their specific public policy involvement. That information has been best detailed in case studies. Hence, very little is known about which organizations actively represent particular interests, issues, and policies.

Even the categorizations of interest groups explain little of policy relevance. Groups usually are typed according to membership, as, for example, farmers or agribusinesses. While this classification provides useful information about such matters as the percentage of the agricultural lobby representing farmers or trade associations, the data do not provide insights into who wants what from government. Farmers and the organizations they select to represent them are not all alike, nor are they directed toward the same public-policy ends. Corporations, trade associations, unions, consumer groups, and other organizations are all subject to diversity in their policy preferences.

Confronted with numerous implications of farm, food, and foreign issues, agricultural groups should be expected to have not only diverse but fragmented policy concerns. New organizations, to whatever degree they exist, are likely to represent new demands affecting the agricultural sector. Have these organizations, through an expansion in representation, been the ones responsible for what Don Paarlberg described as the USDA's new agenda of nontraditional farm policies?[14] Or have established organizations developed new interests and helped bring about a broader agricultural policy? Answers can be found only when the relationships between interest groups and public policy are better understood. Once answers are discovered to "Who wants what?" the forces of both change and resistance can be identified. It will then be possible to identify whether new policy claims are seriously challenging past farm policy practices or merely adding new but independent dimensions to traditional farm maintenance policies.

WHAT ARE THE EFFECTS OF AGRICULTURAL CHANGE ON INTEREST REPRESENTATION?

Agriculture itself has been in great flux. Not only have other components of the food system established their economic importance relative to farming, but producers and agribusinesses have all had to confront conditions that have made them more vulnerable to worldwide market conditions.

Farms are larger, more capital intensive, and forced to operate according to sound business practices. Many producers have been failing financially since the late 1970s. The world food market offers competition to American producers and traders from countries that only a few years ago were dependent on imports. Increasing biotechnological breakthroughs continue to add to international commodity surpluses. In addition to overproduction, farmers and businesses have been troubled by accelerating costs, high interest rates, national budget deficits, the high value of American currency, protectionist intervention in international food trade, and declining land prices. The threat of more farmers and greater amounts of land leaving agriculture is real for both the agribusinesses and rural communities that are economically dependent upon a highly productive agriculture.

This situation fails to replicate past cycles of agricultural booms and busts. Farmers, their financiers, and their suppliers are far removed from those times when the prevailing wisdom suggested temporary belt tightening and marginal adjustments while waiting for better conditions. For as many as 214,000 American farms, economic revival may never come.[15] International observers advise that commodity prices must become more competitive.[16] That means lower, and lower prices translate into diminishing opportunities for generating the increased income that would bail out debt-plagued farmers.

To what extent has the agricultural lobby responded to these changes? Are organizations attending to all those issues and policies that address a transitional agriculture? Or are groups selectively attentive, more wedded to past policy priorities?

Political scientists have never agreed whether lobbying is principally a response to emerging issues in the political environment or a reflection of the biases of activist elites. To what are agricultural lobbyists responding? The obvious interests of financial supporters? Short- or long-term supporter interests? Personal policy preferences of staff and officials? Maintaining a place of prominence for themselves or their organizations in the agricultural policy process? The less attentive organized interests are to the long-term policy needs of their constituents, the less possible it may be for a group or firm to become a participant in policy deliberations.

DOES THE STRUCTURE OF ORGANIZATIONS AFFECT POLITICAL REPRESENTATION?

Increased lobbying success has been linked with a narrowing of political demands.[17] James T. Bonnen has argued that the intensive policy focus of the commodity groups displaced in importance the more diffuse demands of the general farm organizations.[18] The heterogeneous memberships of the Farm Bureau and the Farmers Union apparently produce a variety of viewpoints, competing interests, a larger number of issues, and greater difficulty in compromise. Corn farmers or wheat growers can more easily coalesce around basic commodity programs. Is a narrowing of specific policy preferences related to such structural characteristics of interest groups as who belongs, how an interest is organized, and how autonomous lobbyists are?

If this relationship does exist, are changes in the agricultural lobby sweeping ones or are they simply redefinitions of who represents what? A number of protest-style organizations have arrived on the political scene with comprehensive reform interests. Consulting firms are usually noted for their narrow and technical expertise on agricultural issues. Have these and other developments among organized interests altered the policy process, the requirements for effective representation, or the expectations policymakers have regarding credible presentations by lobbyists? Has this meant that some organizations have a newly found or recently elevated capacity to influence some important aspect of agricultural policy? If so, are there any policy biases in the kinds of issues and agricultural problems that are not effectively represented?

DOES INTEREST REPRESENTATION HAVE AN OBSERVABLE IMPACT ON AGRICULTURAL POLICY?

Political scientists usually look at influence—that is, the ability of one policy player to move a government decision in the direction that this player desires—as a factor derived from the presence or paucity of resources. Better strategy, more money, greater lobbyist skills, membership characteristics, and the like have been seen as the determinants of an organization's political influence by nearly all interest group observers since David Truman. The emphasis of this orientation has produced a peculiar but hardly surprising finding in an area of research in which systematic data are hard to collect. That is, some organized interests have measurably better resources than do others. Sometimes these more resourceful groups win, but often they lose. So, too, do the less resourceful. No one, however, has demonstrated with what frequency this occurs. The conclusion that sometimes groups matter and sometimes they do not has been easily drawn and attributed to the complexities of a policy process populated by many participants other than interest groups. Since it attributes success to little more than chance, this summation of group influence as an occasional phenomenon is unsettling. Interest groups may matter, but, as Bauer, de Sola Pool, and Dexter once asked, how often are they likely to influence policy, and under what circumstances?[19]

Studying an array of interrelated organizations in an arena in which many policies and issues are always being decided, and in which not all groups can be involved in all decisions presents a unique opportunity to look for group success and get some measure of its frequency. One need only look at policy decisions as independent variables and group involvement in the decisions and their positions as dependent upon them. Measures of successful participation can be determined, as can the relative success of different groups and types of interests. Such data can provide numerous insights into how effectively interests are represented by the agricultural lobby. An analysis of a sample of policy decisions, with an emphasis on the question, Who promoted what proposals? also provides an opportunity to judge the degree of involvement of interest groups compared with that of participants in the agricultural policy process.

HAVE COALITIONS OF INTERESTS BECOME ESSENTIAL TO AGRICULTURAL POLICYMAKING?

All of the omnibus farm bills of the 1970s and 1980s have had important coalition bases.[20] The legislation was passed after obtaining support from congressional members with different sorts of constituents. In the final passage of the last four farm bills, it is believed that labor, urban interests, conservationists, and other nonagriculturalists exerted a powerful influence. Provisions were developed and included in each instance specifically to attract such outsiders.

Because of the dearth of policy data, political scientists know little more about coalitions than that lobbyists use them to help one another and to resolve conflicts. Influence has only been presumed for interest groups relative to congressional coalitions, an organizational form that has been studied extensively. Despite considerable intergroup coalition building by organized interests, no one has examined whether interest-based coalitions help bring about congressional coalitions. Given group ties to locally active and voting members, there might be a constituency bias linking interest and congressional coalitions. Does this exist, and if so, does the need to forge congressional alliances account for seemingly unrelated interests' working together to agree on and maintain policy compromises? The large number of both ongoing and ad hoc coalitions that meet on selected agricultural policy topics suggests that these alliances provide some useful purpose in policymaking.

What distinguishes those agricultural issues around which coalitions form? Are coalitions less active on narrow, technical matters? Are they unlikely to exist in the absence of conflict? Or do they just serve a sorting-out process that informs the participants but results in few policy consequences? Clearly, there can be no comprehensive understanding of the agricultural lobby, what its different elements represent, and their policy impact without discovering why these organizations so frequently work together.

DOES THE FARM BILL INFLUENCE AGRICULTURAL POLICY?

Certainly the renewable farm bills, which are based on permanent legislation and to which agriculture reverts in the absence of periodic reconsiderations, sets forth and directs much of agricultural policy. Farm, food, and foreign policy issues all find their way to each farm bill. With commodity provisions, Food Stamps, Food for Peace, and export assistance programs all being renewed every four to five years in that act, the farm bill is the unchallenged center stage for most agricultural interests.

Does that center stage and the way in which legislative provisions are structured in a single omnibus act affect the policymaking process, especially the representation of interests? Commodity programs are considered one at a time, each with its own unique procedures, and they often originate from different subcommittees. Domestic food assistance and international food aid are also unique and handled apart from basic farm provisions. Specific provisions can be amended to these bills. Is it this structure and its attendant advantages that have brought commodity groups and other intensively organized interests a reputation for exceptional influence?

Of course, it is impossible to research agricultural policy removed from the potential influence of farm bills, but important points of comparison can be made. For instance, how do those diffuse interests that want several things in any single act fare? Do organizations that want to challenge or change provisions either gain access to subcommittee members or succeed with their policy priorities?

There is always the opportunity of comparing farm bill legislation with other legislative components of agriculture policy. Each year Congress considers and passes other related pieces of legislation that lack the disjointed structure of major farm bills. When involved, are the commodity groups equally successful on these bills? How does the influence of other organizations compare in this different context? If an assessment of the relative influence of agricultural lobbies is to be made, the structural consequences of legislation must be considered as a control variable.

Those questions should make it clear that this book is as much about agricultural policy as it is about agricultural interests. Although much more remains to be learned about lobbying and group behavior in general, the policy dimension of that behavior is the most neglected. The contribution this book aims to make could be summarized as answers to two final questions. How do interest groups behave in determining their specific policy priorities? How does this behavior relate to an organization's chances of winning?

THE FOLLOWING CHAPTERS

The material in this book combines the 1985 survey data discussed in the preface, some previously collected responses from earlier research, and relevant historical information. The intent is to focus on each of those questions that have provided the impetus for this book while providing a readable account of an agricultural lobby in transition. Readers will not find the following chapters to be an encyclopedic description of a collection of groups, however. The purpose is to generalize from the data and provide an analysis of what this collective lobby and related private interests mean to agricultural policy and policymaking. For this reason, there are few tables, and data are reported on the basis of the respondents' prevailing beliefs, rather than in terms of percentages. The intention is to let group representatives speak for themselves as much as possible.

Chapter 2 provides an overview of private agricultural interests and how they are organized around the basic components of the food system. Representational changes, as they have occurred over the past seventy years, are also addressed in that chapter. Chapter 3 looks at the ways in which the agricultural lobby operates and is currently organized. Organizations and their public policy activities are looked at in the context of a broadening continuum of intensive versus diffusely involved organizations. Four types of interests are identified: agrarian protest groups, multipurpose organizations, single-issue organizations, and single-project organizations.

These interests, as they have evolved historically, provide the topics of the next several chapters. Chapter 4 looks at the emergence and limited institutionalization of farm protesters and their nonagricultural supporters. Chapters 5, 6, and 7 compare the multipurpose and single-project organizations. Many groups and firms, for reasons of organizational diversification, have moved from rather narrow to necessarily more broadly based policy interests. These chapters focus on farm, agribusiness, and public interest lobbies. Lobbyist consultants, attorney representatives, public relations specialists, economic analysts, and research institutions—each with the capacity to approach very narrow assignments with intense concentration and highly specialized expertise—are covered in Chapter 8.

Coalitions politics is the subject of Chapter 9. Chapter 10 reviews patterns of representation in terms of important policy implications of who does and does not represent what. The politics of the Food Security Act of 1985 is specifically analyzed in Chapter 11. The concluding chapter summarizes the book's findings within the context of interest group theory.

The data suggest that the agricultural lobby is best characterized by a diversity of groups that attend, largely on their own, to unique policy interests. In doing so however, these organizations leave a surprising number of issues to be influenced by legislators, administrators, and other participants in the policy process. Part of the reason for this rests with the complexity and constant change within the agricultural lobby, the dynamics of which are examined in the next chapter.

2

The Agricultural Lobby

The agricultural lobby, like other human enterprises, is constantly changing. Sometimes human efforts succeed; at other times they fail. Some things are done better than others. Some activities run their course as goals are attained. Others cease as participants lose enthusiasm, turn elsewhere, or redirect their loyalties. Resources are exhausted, used up.

Interest group theory provides a cogent and precise explanation for the dynamics of change within the agricultural lobby. Most organizational efforts have been nurtured by entrepreneurial investment, by the leadership activities of one or a few individuals who define and create a marketable plan.[1] Entrepreneurs, however, have depended on the support and goodwill of those who become the organization's supportive patrons.[2] As long as transactions between entrepreneurs and patrons work, organizations can emphasize external relationships in pursuing political influence for themselves. When the sustaining relationships cease to produce satisfactory arrangements, the attendant adjustments usually range from redefinitions of member services to basic redefinitions of policy interest. At the extreme, organizations wither and disappear.

EARLY POLICY CONTEXT

The history of American agriculture has shown a strong tendency to link new organizations with public policy demands. Farmer rebellions and the protest groups they produced were the earliest attempts by agriculturalists to influence government. Typically these organizations emerged when farm incomes were comparatively low or when price fluctuations caused major price failures. Each of the many groups that gained national or regional prominence, even when organized primarily around the social benefits of agrarian fraternity, actively sought political answers for members' problems. Membership growth and increasing politicization of the organization went hand in hand for most of them.[3]

These early agricultural groups also demonstrated clearly established political interests, which, while subject to changing conditions, gave distinct and even ideological identities to each organization as it sought to bring together both individuals and local farm groups.[4] For example, the National Grange will be forever associated with its stands against the railroad and for government regulation in the 1870s. The National Farmers' Alliance, which was organized in competition, attacked the Grange for a lack of militancy and urged the adoption of farm cooperatives as a more assertive action.

When the National Farmers Union (NFU) came forward in the early 1900s from the ranks of the collapsed Alliance, it carried the cooperative movement one step further in urging central marketing institutions. Its contemporary, the American Society of Equity, argued against the NFU for free market rather than anticaptialist reforms. Equity wanted farmers to withhold commodities and control the market by emulating big trusts. This organizational give-and-take characterized the relatively few nationally oriented groups that made up the farm lobby in its earliest form.

As federal farm policy moved from regulatory decisions that established social policies protective of farmers to policies of more direct support of and involvement with selected types of producers, what had been an exclusively grassroots agricultural lobby slowly moved to Washington. Beginning in the early years of the twentieth century, agrarian protest movements with their goals of electorally based reform gradually gave way to the establishment of Washington offices. By the 1920s, the agricultural lobby was still exceptionally small by today's standards and was just beginning to have a strong relationship with Washington policymakers.[5] The four national organizations that first came to Washington—the American Farm Bureau Federation (AFBF), the National Board of Farm Organizations, the Farmers National Council, and the Grange—represented points of view that were scattered among an immense number of local and regional groups of cooperative, farm defense, social, and self-help organizations. These estimated 8,600 groups were attractive and necessary targets for any Washington organization that wanted to claim a national following as a potential power base.[6]

Mining these locally dispersed and skeptical resources was impossible for national organizations that could not justify farmer affiliation and support. With a scarcity of personal services to provide, each of the four national groups had little choice other than to emphasize the uniqueness of its public policy positions and to claim a political high ground. Although small, the agricultural lobby represented a complexity of policy choices almost from its arrival in Washington.

The Farm Bureau stressed its ability to secure congressional support for a variety of farm needs from fertilizer supplies to processor regulation and technical assistance. The National Board, with its strong ties to the National Farmers Union, was more concerned with cooperatives and collective farmer action. As a consequence, the board initially viewed the government more as a facilitator of agriculture than as a direct contributor to it.

The other two national organizations grew out of internal Grange conflicts. Several progressive Granges, in defiance of National Grange policy, had become part of the Nonpartisan League and related farm-labor coalitions. The Farmers National Council became the Washington representative of the progressive Granges and advocated such radical agricultural policies as nationalization of the railroads. Had it not been for the council's presence in Washington, the Grange itself probably would have stayed away. Its primary goal was to repudiate council positions and generally articulate the position that farm solidarity was more important than government for agricultural prosperity.

As the interaction between the nonradical agricultural lobbies and policymakers settled into the comfortable arrangements that followed the social reform successes of the congressional Farm Bloc, interest representation hit a policy high with an attendant political low for effective agricultural pluralism. The Farm Bloc, short-lived as a formal coalition in 1921, provided the means for farm interests to begin a long and friendly relationship with legislators, who for nearly three decades thereafter used the organizations to communicate and gain favor with farmer members.[7] Under these circumstances, the three aggressive farm organizations—the Farm Bureau; the Farmers Union, which soon supplanted the National Board; and later, the more reticent Grange—were able to cultivate responsiveness from agriculture committees, who, as an informal extension of the 1921 Farm Bloc, continued to hold sway over a generally supportive Congress.

The political low to these congenial relationships concerned their impact upon other agricultural organizations. It seemed to both policymakers and many later scholars that the general farm organizations that represented numerous political issues were the only components of the agricultural lobby and the only organizations in agriculture during this period of severe and prolonged farm depression. In fact, they were not. Congress simply did not want to listen to radical views on such matters as market intervention, so such groups as the Farmers National Council were forgotten, even though similar local organizations maintained an intense grassroots activism in the Midwest. It was just difficult to hear them. In an era when travel and interaction were difficult, members of Congress lacked the means to establish a dialogue with the thousands of local organizations, so congressmen focused their attention on the political mainstream as represented in Washington by group leaders who could see them regularly.

There were also some organizations of both agriculturally dependent businesses and single-crop producers. Wanting redistributive public policy, both types articulated policy positions of their own but did so without the same Washington presence or cordial congressional reception as most of their general farm counterparts. Agribusiness lobbies, particularly the farmer-mobilizing American Council of Agriculture and those in the implement industry, argued for market supports to enhance farm income and, incidentally, sales of farm equipment. As commodity prices plunged in the early 1920s, producer associations of dairy, wheat, soybean, cotton, tobacco, cattle, and other commodities clamored for similar government action to force price increases.

The success of general farm organization, though often couched in terms of victories over robber barons, was really a victory of one organized interest over another. The social reform activities of the Farm Bloc, which were focused on regulatory control, were directed against middleman industries that had organized to gain competitive advantages over their farmer suppliers. Several agriculture lobbies of the 1980s were among them, including the North American Export Grain Association, the Milk Industry Foundation, the Corn Refiners Association, the American Seed Trade Association, the American Meat Institute, the American Feed Manufacturers' Association, and various boards of trade.

This single set of regulatory successes marked the beginning of the modern era in agricultural lobbying. While there was real diversity of interests and policy preferences, lobbying techniques pioneered mainly by the American Farm Bureau Federation obliterated the impact of nearly all the others. Using its massive Midwest membership, the Farm Bureau's Washington office brought grassroots discontent to Congress by informing and activating constituents both at home and in Washington. Farm Bureau lobbyists insisted on congressional accountability for policy decisions, gathered and distributed sophisticated data on farm conditions, organized discussions between legislators, and kept visible on Capitol Hill.

With the Farm Bureau's potential for organized electoral connections, congressmen found it increasingly difficult to respond to the rest of the agricultural lobby. Because the claims of locally affiliated farmers were so fragmented, it was hard to comprehend what would satisfy them. In the face of opposition from the general farm groups, market intervention seemed too risky. Orchestrated opposition even dulled the effects of the back-home lobbying of agricultural middlemen. What seemed like a Farm Bureau steamroller soon caused other components of the agricultural lobby to join with that organization if they wanted to advance their own interests.[8] Some organizations simply left the policy arena, emphasizing instead how best to cope with government regulation and still keep profits high.

Although the Farm Bureau soon became a convert to farm-maintenance pricing policies and was frequently responsive to the initiatives of other organized interests, agricultural pluralism did not return to the Washington lobby in the three decades after the early 1920s. Even while policy adjustments were made in market approaches to agricultural programs throughout the period, most decisions were in-house products of the agricultural establishment or policy system.[9]

This trend was accelerated as government regulation of agriculture expanded from the social reform of earlier eras to include greater administrative discretion over the allocation of benefits from farm programs. While the Farmers Union became more a voice for smaller-scale farmers, policy differences between the general farm organizations were never especially sharp in Washington during this time. The greatest evidence of diversity of interests within the agricultural lobby was manifested by the grassroots protests of the Farm Holiday Association in the 1930s and the National Farmers Organization, created in 1955. Because of their harsh criticisms of farm interest representatives, both organizations were branded with the same sweeping radical label that had been attached earlier to such diverse groups as the Farmers' Alliance, Equity, and the Farmers National Council. This labeling blunted the policy emphasis of both organizations and further reduced their credibility. As a consequence, commodities withholding by producers and collective bargaining never gained enough Washington-based policy advocates to equal or surpass those of their counterparts in organized labor.

THE NONPOLICY COMPONENT

To acknowledge the linkage between specific policy interests and the emergence of the organizations does not establish a causal relationship between the two. It is not sufficient to say that position taking on public policy will lead to an organized response. Many ideas generate neither an attempt to organize nor a following. Nor is it necessary for an emerging organization to take a public-policy stance. That point becomes critical in understanding the agricultural lobby. The number of distinct organizations from the private sector representing members who share an interest is truly enormous. There are at least 1,900 state and national groups and firms organized around nonfishing agricultural interests that are similar to those active within the national agricultural lobby.[10] Only 440 of these organizations undertook the nominally active task of testifying in legislative hearings during agricultural policy debates between 1977 and 1979.[11] Fewer than 300 of them acknowledged employing Washington representatives during the farm bill proceedings in 1985.[12]

These numbers reveal the limited importance most agricultural organizations attribute to politics and public policy. Political action is by no means the central feature of their organizational lives. Except for a few organizations, most of what could be collectively identified as the agricultural lobby comes and goes.

This coming and going cannot be interpreted as a result of a modern political age in which interests simply hire political representation from a professional Washington lobby. It must be remembered that the historical existence of locally organized farm groups centered around problems of rural isolation. The 8,600 local groups of the 1920s were targets for the national organizing of the early farm lobby rather than vehicles that promoted a Washington presence. It must also be remembered that the organization of groups by commodity was evident by the first two decades of the twentieth century. Although commodity representatives made considerable policy noise in 1921 and 1922, by the early 1950s only one commodity organization was considered an independently active political force.[13] Wheat, cotton, and corn growers had to gain whatever policy influence they could through the general farm groups or through personal relationships between wealthy individual farmers and key legislators. Only fifteen years later, though, nationally organized groups representing each major commodity joined the National Cotton Council in visibly dominating farm legislation.[14] Over half of the dozen or so commodity groups nationally active by the late 1950s were thirty years old or older. Some dated from the turn of the century. Of the remaining "new" lobbies, only pork and sorghum interests were not clearly identified with existing regional commodity groups.

Why are organizations formed? For the agricultural lobby, the answer is at least partially explained by the non–public policy activities that presently engage most of them. In fact, it makes more sense for organizations to structure incentives for supporting a potential lobby around nonpolicy activities than around activities related only to policy ends. As Mancur Olson first observed, successful efforts to provide public policy benefits normally bring about collectively received goods or rewards.[15] Even when a legislative victory distributes highly specific or selectively assigned rewards, these benefits go to all those who share the characteristics that define the recipients. Organizational supporters who give of their time and money will benefit no more than will those who save by not joining. Establishing or maintaining an organization on the basis of lobbying potential alone is clearly a risky business.

Entrepreneurs who hope to nurture organizations that may act as successful interests need to minimize this risk, so these individuals lobby to satisfy those who want collective action and also to provide several kinds of selectively received goods and services that can by obtained only through membership. Robert H. Salisbury categorized these member incentives as material, solidary, and expressive benefits.[16]

Over a very broad range of time all three types of benefits have been found to be important for the farm components of the agricultural lobby.[17] The National Farmers Union operated grain terminals and cooperatives that offered material savings to its members. These, in turn, became the initial focus of its Washington lobbying and its efforts to target new local supporters. The Grange was organized around social interaction and agrarian fraternity. To preserve the solidarity that sustained strong local ties, the Grange, in contrast to the Farmers Union, limited its involvement in policy. While emerging, the Farm Holiday Association, National Farmers Organization, and American Agriculture Movement each shared the tendency to forsake strategic policy advantages when they interfered with the needs felt by farmer protesters to attack institutionalized policymakers.

These and other examples suggest that selective incentives are more than just supplements to political action. Member incentives and policy activism appear to be directly related. Indeed, in many instances services structure policy responses. It hardly seems an accident that member-based lobbying and close relationships to government were pioneered by the Farm Bureau, an organization created by federal funds to develop a mass membership that could be encouraged to increase agricultural production to satisfy the consumption needs of World War I. Nor was it an accident that the agribusiness trade associations of that era were not prepared with a similar lobbying response. The services of the trade associations revolved around integrating each single-interest industry in a clublike atmosphere that slowly gave way to efforts to fix prices.[18] When attacked by the newly aggressive farm lobby, the associations had no basis for generating a proactive plan for government intervention for legitimate price stability. The associations had always accomplished their ends through their own internal means, without government. The only conceivable approach was to fight regulation rather than to accept a new political order in which government as a virtual stranger was called on to help.

Viewing member services as consequential to the behavior of an organization explains three important aspects of agricultural lobbying. First, for its potential to be so vast, the dimensions of what an organized interest undertakes must be manageable. Players come and go, depending on whether the organizational activities provide a point of departure for getting involved in public policymaking. Second, involvement is somewhat predictable. Environmental changes, the rise of new issues in policy circles, perceptions of changing member needs, or new organizational staff with new service responsibilities may trigger political involvement among previously uninvolved participants. Reversals in these conditions may signal an end to participation. The need to address a new issue will also affect the degree to which unresolved policy concerns receive attention. Organizational resources, after all, are not unlimited and must expand from a base of member services without detracting from them. Third, the shared service/policy orientation explains why agricultural organizations appear in waves. The 1950s were marked by more than the milk-dumping protests of the National Farmers Organization. In those years the political transformation of commodity groups began. David Truman saw waves of new activity as typical of American interest groups.[19] If agriculture is similar to other policy areas in the degree to which organizations already exist, Truman probably witnessed newly activated or reorganized interests, rather than all newly formed ones.

The conditions of the post–World War II era had shifted by the early 1950s. As a result of consolidations and the closing of many smaller units, farms were larger. Farm production was increasingly specialized by commodity. Efficiency and technology created by more capital-intensive agriculture were producing more surplus and greater political pressures to reduce support for basic crops. Existing commodity organizations, along with large producers, were more involved in crop- and animal-specific service programs, especially through the land grant colleges and the USDA's Extension Service. Organizations could no longer satisfy members by concentrating on disease problems, information exchange, market assistance, and lost or stolen animals.

Not all commodity groups used this time to get equally involved in promoting price support policy, however. Both the National Association of Wheat Growers and the National Corn Growers Association were founded in the 1950s, but the Corn Growers did not establish a Washington office and seek ongoing political visibility until the 1980s. Despite years of legislative cooperation between the two groups, the Wheat Growers had been advocates of higher supports and more government involvement than had the Corn Growers. The National Broiler Council was also a product of this decade. It represented a technology that effectively enhanced demand, though, and the long-standing free-market approach of its members became even more strident during this period. Broiler Council activism was in no small part motivated by the general recognition that feed prices were now threatened by wheat and corn producer lobbying as well as by the nongovernmental factors that had been monitored by the predecessor organization since the 1930s on behalf of broiler producers. Reorganization provided a national group structure better equipped to perform as the loyal opposition in commodity politics while actively engaged in traditional services.

The next wave of additions to the agricultural lobby arrived under similar circumstances. Of the organizations most active on the 1981 and 1985 farm bills, over one-third were founded between 1969 and 1979. While some of them originated through mergers of trade associations, and others were newly opened consulting firms, most of them were involved with recent issues. About half of the new groups represented the "new politics" of consumers, food needs of the world, and such domestic issues as rural development and the role of women. Most of these public interest groups articulated the traditional regulatory goals of social reform, although in this instance reform was directed against agricultural practices rather than for them. Unlike previous additions to the agricultural lobby, most of these were not mass-membership groups. Instead, their support came from other interest groups, charitable organizations, professional associations, or foundations. In short, patrons helped make possible new agricultural lobbies to extend the work of ongoing organizations.

The respondents to the 1985–86 survey shared this orientation. All of the organizations, even modern farm protest groups, were organized around selectively provided services. Public policy activism, first and foremost, means performing as watchdogs who study the policy arena for signs of needed intervention by the group. Washington-based representatives uniformly acknowledged the need to conserve resources in deciding which policy decisions to address. Each organization had to assign rigid priorities to issues. Not even the exceptionally well-staffed Farm Bureau and Food Marketing Institute addressed all agricultural issues of immediate, let alone potential, interest to members. When asked about the most important criterion, lobbyists commonly referred to the need to protect those interests that members were most accustomed to getting from the organization, whether they be service or policy benefits.

Policy vigilance—watchdog politics—was emphasized over policy demands because, in what was seen as a rapidly changing agricultural arena and a consequently uncertain policy process, threats were everywhere. For most of the mass-membership groups and all of the organizations that survived through contracts and grants, uncertainty produces an increasing directness in dealing with members and patrons who are worried about political events. Some respondent groups lobbied less, but more had cut back on member services as they sought to reallocate resources and thus satisfy financial contributors.

Nearly two-thirds of the respondent organizations mentioned strategies for coping with perceived changes. Most commonly, this entailed adding new staff with new skills or experience in public relations, technical and personal services, or previously neglected issue areas. It also meant a more vigorous search for ways to increase members' involvement in lobbying as well as for new members and other sources of funding.

Organization building of this kind was felt to be widespread by agricultural policy activists. The respondents of all but two organizations agreed that there had been considerable growth in the number of involved organizations over the last several years. Many termed this growth dramatic and startling. In other words, there was a perception among agricultural lobbyists that helped explain David Truman's wave phenomenon on the basis of more than just policy disequilibrium. Not only were organizations responding to change, but they were creating changes as well. Most of these changes were internal ones that affected staffing and member relations. Internal modifications, however, were often designed to have consequences for national politics or public policies that would affect other groups and firms as well. Depending on what resources were available and what opportunities were seen by organization staff and officials, the organizations concerned frequently altered the extent of their own participation. At least 92 of the 130 respondent organizations reported that their involvement in agricultural policy matters had somehow increased since 1975. Nonetheless, several felt compelled to reduce their involvement, with a few organizations all but dropping out of politics. These data and opinions suggest that changing political conditions with their disrupting personal effects are perceived as opportunities for impressing on those who support the organization financially the need for collective action.

LOBBYING AND THE FOOD AND FIBER SYSTEM

The preceding sections outlined the historical changes in the agricultural lobby as it evolved from farmer movement to diversity in representation. The lessons of the past indicate that continuing change is the most predictable feature of that lobby. One should never presume too much about what organizational changes the future will bring. It would have been difficult to predict, for example, that the American Farm Bureau Federation would emerge as dominant in a highly pluralistic set of Washington lobbies. The Farm Bureau lacked the one leader-and-member feature common to all earlier nationally prominent farm organizations, a movement style based on policy-related reform. Nonetheless, the AFBF gained its political strength on the basis of membership relations; and it later used those diverse farmer members to quell representational pluralism in agricultural policymaking. It was no less unpredictable—and ironic—that AFBF's policy influence would be lost, not to partisan conflict and another giant farm interest, but rather to a coterie of much smaller and individually less resourceful groups whose members had operated quite frequently under the AFBF's protective cloak for many years.[20]

The reasons for organizational uncertainty and instability have been briefly outlined in terms of the universe of bifurcated expectations in which agricultural interests exist. In essence, interest groups are expected to show two faces to two different constituencies. One is a policy face. Agricultural interests, after all, have achieved their organizational importance in a policy context, not as social clubs or service providers. Policy-related expectations are especially great when groups and firms headquartered in Washington hire lobbyists or specifically organize around issues of political change. Comments from policymakers and from representatives within the agricultural lobby assess organizations in terms of such factors as their lobbyists' skills, organizational effectiveness in congressional relations, or ability to raise Political Action Committee (PAC) funds. Merit is awarded on the basis of how well a participant plays the policy game, not on the basis of whether the game needs to be played. Profits may be high. An industry may be in financial order. A firm may be without a policy motive. Washington insiders, however, still find fault for lack of both effort and foresight. The same evaluative tendencies apply to non-Washington groups that take a policy interest. Organizations that lack similar involvement are seen as irresponsible, useless, or dangerous, no matter how well they service the business or personal needs of farmers or businesses in trouble.

The service face of an interest group is one that, while necessarily worn for the members, brings it no glory from the viewpoint of the Washington representative or the policymaker. It only pays the bills. While organizations would lose support if they did not lobby, most groups would find membership losses more severe if their service efforts failed.[21] As a consequence, organizational entrepreneurs—who are often their interest's managers rather than lobbyists—pay considerable attention to the members' service needs, often structuring the group's identity around them. Lobbying, when it is necessary, usually results from an extension of an organization's broadly defined purpose, including services, rather than from a desire to be the best lobby in Washington. When resources are scarce, policy initiatives are especially likely to be derived from services. Groups move politically to protect what they know members value. If more resources are available, or the organization is more capable of raising revenue from well-financed patrons, then lobbyists can pursue a wider range of policies and issues. The perceptions of scarcity of time and money among representatives of agricultural lobbies indicate that their policy involvement seldom takes them far from the heartland of their group's primary, member-related identity. Lobbyists feel that they have little time for philosophical issues, ideological confrontations, programs that their members do not use, proposals that are liable to lose, anything that will convey a negative impression of the organization, or issues that are difficult to understand but that need to be sold to policymakers. So, although being an organizational representative is strongly associated with a total involvement in the drama of policymaking, lobbying reality is often harshly restricted by concerns over what expended efforts actually bring to the organization. For agribusiness firms, whose executive officers must be concerned with profit maximization, lobbying is especially hard to justify.

This relationship between policy and nonpolicy concerns of organized interests must be carefully considered in attempting to understand who represents what in agriculture. Because of the varying priorities that different organizations accord lobbying, criticisms of agricultural interests abound. Lobbying in agriculture is often said by policymakers to be uneven, biased, inconsistent, hit-or-miss, inattentive to detail, unconcerned with major issues, and too incremental. These criticisms are little more than a reflection of that collective lobby's inability to provide equal and practical consideration to the wide range of farm, food, and foreign policy issues relevant to agriculture. This criticism can best be explained by looking at American agriculture as a complex food and fiber system in which commodities are produced on farms and eventually brought to consumers. Interest representation seems to fall heavily to production issues, however.

Tom Veblen has developed a systems approach for the United States food industry that is also applicable for fiber and tobacco products. His model can be used to develop an understanding of what issues and problems account for the variety of groups mentioned in Chapter 1. The food system, as Veblen portrays it (see Figure 2.1), begins with various animal and other agricultural producers who turn their commodities over to merchants. Merchants, however, are not buying from U.S. producers alone. Foreign producers who export to U.S. markets are also shown. Their competitive status is represented by their place alongside U.S. producers.

The merchants are only the first of a series of industrial middlemen who handle farm commodities before consumers receive them. The chain of gray arrows reveals the diversification among agribusiness middlemen. Processors buy from merchants and then sell to manufacturers. Value, in the form of increased worth with attendant consumer cost, is added at each link in the chain. After manufacture, food goods go to distributors, who send them to grocers; to food prepa-

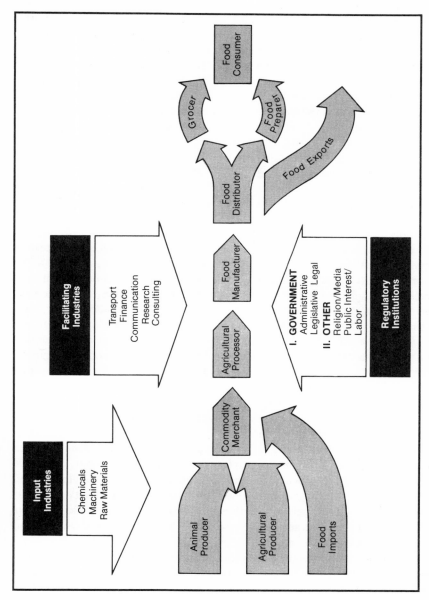

Figure 2.1 Food Systems Model for the United States. Reprinted by permission of Food Systems Associates, Inc., Washington, D.C.

ration businesses, such as restaurants; or to exporters. Food consumers, in the United States or abroad, eventually receive what farmers produce. There may be six domestic ownership transfers before goods are received, however. Exported food may go through one to three other ownership transfers. Some export sales go to foreign governments and then to consumers. Others go first to commercial importers, then to distributors if import merchants do not distribute them, and, finally, to retail consumer outlets. Each of these sets of business transactions represents more than just different handlers and potential owners. In the strictest sense, the transactions also create distinct interests. Farmers want to sell high; each middleman wants to buy low and then sell high; and consumers want to buy low. There may be other values attached to food goods that affect and further confuse definitions of interest, of course.[22] Among other noneconomic preferences, farmers may associate quality-of-life interests with products and consumers will, to some extent, prefer variety and nutrition to low cost.

For the agricultural lobby, the diversity of interests in the food system chain has become institutionally organized. Each link has come to be represented by associations of producers, handlers, and consumers that serve members who join each organization. Not every organization has responded by making public policy representation important. Several of the groups that have emerged are active lobbies, some are infrequent lobbies, and still others are only latent or potential lobbies.[23] Not all potential food system interests, in other words, have representational status in government decisions about who will get what benefits of agricultural policy.

In addition, not all of the organizations that have shown policy involvement define their interests as equally broad or narrow. The Veblen food system model is again useful in explaining this point. Food manufacturers, for example, may focus on public policy decisions that affect food prices relative only to their direct relationships with processors who sell to them and distributors who buy from them. Alternatively, manufacturers may take a much broader policy view—or interest—and focus on political decisions up and down the food chain. This approach would bring them into decisions about farm price supports that increase food costs as well as consumer food safety issues that have the same pricing effect. Clearly, those who decide organizational lobbying tactics and approaches within groups and firms have a great deal of latitude in their choices. But, as noted earlier, because of internal organizational matters, their policy options are not unlimited. For middleman agribusinesses, policy choice limitations are restrained further by another form of diversification. Many firms, and thus many members of trade associations, are conglomerate corporations with subsidiaries of more than one type. As a consequence, their interests and policy involvement are even more difficult for the organization to determine.

Other factors create an impact, too. The two black and white arrows at the top of Figure 2.1 show agricultural input and support industries. Input industries, for the most part, provide products required for modern, capital-intensive agri-

culture. These industries include chemical companies, farm equipment manufactures, and a variety of firms providing fertilizer, plants and seeds, and other raw
materials. For much of agriculture—especially in the vegetable, fruit, and specialty crops—labor remains a vital and high-cost input. Each of the input industries is organized and, at least to some extent, active as an agricultural lobby.

Support industries—or, as Veblen calls them, facilitating industries—are
more varied than input organizations; and, while organized, they are generally
less active as agricultural lobbies. Many of them, in fact, are involved in so many
sectors of the economy that their own interests in agriculture are not clearly defined. Most support industries serve both farms and agribusiness industries, although not in any integrated way. Major facilitators include railroads, truckers,
and the maritime industry for transportation; banks and other production lenders; boards of trade; industrial equipment suppliers and packagers; advertising
and communications specialists; land grant colleges and some other research universities; state departments of agriculture and agricultural officials; professional
associations; and foundations, technical analysts, and consultants with specialized agricultural expertise.

The black and white arrow on the bottom of Figure 2.1 represents other private interests that, along with the media, attempt to set standards to affect the
food system and, to a lesser extent, the fiber system. Organized religious groups
are especially prominent. Labor unions play a role. Most of these organizations,
however, are the privately organized public interest groups whose policy intentions are aimed at reforms that will not materially benefit either group members
or activists.[24] Conservation, environmental action, social justice, and animal welfare interests are well represented among these groups. Except for institutionalized food programs, such as school lunches or nutritional programs for the
elderly, most consumer concerns are articulated by those public interest groups
(PIGs). Many of these organizations have been so intently active in agricultural
policy during the past ten to fifteen years that most farm and industry lobbyists
now think of them as yet one more routine factor to be considered in determining
policy planning and related lobbying strategies. Policymakers tend to think of
many of the PIGs as integral parts of the agricultural lobby.

The food systems model offers a great deal of utility in understanding the agricultural lobby. First, it shows the potential for diversity afforded by U.S. agricultural industries. Second, the model helps explain the concept of interest as it
pertains to potential public involvement. Third, potential conflicts that have both
organizational and issue consequences are apparent. Fourth, it demonstrates
that organizational and public policy changes are likely to have reverberating effects that may extend throughout the total industry. This final point is especially
important in that it emphasizes how critical lobbying decisions made by just a single interest can be in disrupting whatever equilibrium exists within the agricultural lobby. One entrepreneur who strikes a responsive chord may well set in
motion a series of events that have important policy consequences for many in-

terests. All of the above, in fact, are characteristic of agriculture in national politics. Together, these factors account for the propensity toward frequent change in the shape and influence of the agriculture lobby. While characteristic, however, such behavior produces a growing but far less extensive agricultural lobby of policy activists than the model—or logic—implies might exist.

WHO MATTERS MOST IN THE 1980s?

One way to judge the dynamics of change in agricultural representation is simple comparison over time. Thirty years ago, Wesley McCune portrayed fewer than twenty-five farm organizations, agribusinesses, facilitating industries, and foundations as the leading policy activists in agriculture.[25] With the possible exception of the National Farmers Union, McCune found their policy participation beset by interlocking arrangements and cooperation as these interests sought to dismantle farm support programs. Agricultural politics in the early to middle 1950s was fairly simple; it was primarily a centralized conflict over maintenance of farmer incomes. The commodity groups that would soon come to forge powerful coalitions in sustaining support programs were dismissed as "other groups and interests."

The politics of the 1980s is dominated by far more players, both in what could best be called "primary" and "supportive" roles. Eighty-four agricultural organizations are seen by close observers of agricultural policymaking as substantively and continuously active in national politics. Not all of these organizations use Washington-based activism and contacts to advance their causes. Most are producer groups, trade associations, and public interests, but several are individual business firms. For the most part, the interests of these organizations in agricultural policy are unique. While, as lobbies, their representatives talk and forge policy-based coalitions, fragmentation of issue concerns are more common to the groups than the ongoing interaction found earlier by McCune.

The remainder of this chapter briefly discusses the most active agricultural lobbies and notes several organizational features that help promote political interests.[26] The next few chapters will further elaborate on these organizations as well as other groups and firms that are of policy consequence but that come in and out of the policy process less routinely.[27]

Organizations active in the 1980s and included in Tables 2.1–2.6 were selected on the basis of several criteria, all of which are somewhat subjective. The first criterion was sustained agricultural policy involvement. Did the organization stay politically active without prolonged absences when issues of perceived interest came up? Second, the organization had to show an aggressive policy interest. Watchdog lobbies that neither personally contacted policymakers nor mobilized members were not included. Writing letters and appearing before House and

Senate committees were considered insufficiently aggressive behavior.[28] Third, the organization had to show assertive lobbying on more than a single issue. Its staff or members also had to get involved repeatedly on recurring issues. This eliminated groups that were involved in agricultural policy only when they were not active elsewhere. Since many important agricultural issues were decided at other times during the 1980s, an organization did not have to be an active partici- pant in either the 1981 or the 1985 farm bill to be included. Lobbies were in- cluded even if their work involved more administrative than legislative contacts. Despite their importance for policy, consultants were not included, since much of their work is conducted on behalf of specific organizations and interests rather than as specific interests. Several of the organizations included were selected only because of their adept use of consultants, however.

As mentioned, active lobbies were divided into "primary" and "supportive" organizations. Forty-six of the groups were given "primary" status; they repre- sented clear and distinct policy positions of their own choosing. These organiza- tions also mounted independent campaigns to influence policies that their officials deemed important. The thirty-eight organizations placed in the "sup- portive" category may have developed their own positions but had reputations for representing them only through coalition-based lobbying. "Supportive" lob- bies lacked financial, membership, or staffing resources to sustain their own lobbying efforts, but nonetheless contributed to organizations with which they worked. Some were organizations whose service responsibilities to their mem- bers consumed nearly all of the resources that otherwise might have been avail- able for lobbying.

Organizations were considered, first, on the basis of my own observations about their behavior on the list of 180 policy decisions used as independent varia- bles for this study. Second, in order to include groups involved in a wider variety of agricultural issues, they were considered when lobbyist respondents specified that either their organization or another met the criteria. Third, groups and firms were also considered if congressional or administrative contacts provided evi- dence that organizations met those criteria. In these instances, both sample and nonsample issues from 1980 to 1985 were used if raised by respondents. Confir- mation by at least two additional lobbyist respondents or policy contacts was re- quired before any organization was added to the final listing.

PRODUCER ORGANIZATIONS

Given policymaker complaints that the agricultural lobby reflects a farm bias, it is somewhat surprising that only 23 percent of the most active organizations are producer groups. There is considerable variety among these groups in policy in- terests, a factor that further confuses the meaning of charges of bias. More than one-third of the groups represent the general farm interests of many different

TABLE 2.1
PRODUCER ORGANIZATIONS

PRIMARY	SUPPORT
American Agriculture Movement (AAM)	American Sugar Beet Growers Association
American Farm Bureau Federation (AFBF)	American Sugar Cane League of the USA
American Soybean Association	Florida Sugar Cane League
Farm crisis committees (National Save	Hawaiian Sugar Planters' Association
the Family Farm Coalition)	National Farmers Organization (NFO)
National Association of Wheat Growers	National Grange
National Cattlemen's Association	Women Involved in Farm Economics
National Corn Growers Association	(WIFE)
National Farmers Union (NFU)	
National Peanut Growers Group	
National Pork Producers Council	
National Wool Growers Association	
United Egg Producers	

commodity producers. Twelve, including four sugar groups, are commodity-specific interests.

In addition, the political viewpoints of the organizations are even more diverse. Of the general farm groups, the American Farm Bureau Federation has long espoused a free-market philosophy of minimal government interference in agriculture. Both the National Farmers Union and the collective bargaining orientation of the National Farmers Organization favor policy intervention by government to enhance producer income. All three organizations have long-established reputations for high-visibility lobbying in Washington.

The American Agriculture Movement (AAM) and the farm crisis committees advocate more basic agricultural reforms and question the value of New Deal farm support programs. Women Involved in Farm Economics (WIFE) occupies a middle-ground policy between the New Dealers and the reformers, borrowing heavily on ideas from each. Like the AAM and the farm crisis committees, WIFE is most involved in generating broad public support for family farmers on a grassroots basis outside of Washington. Its opposition to foreign and corporate agricultural investment is integral to its public plea and gives WIFE a relatively distinctive activist identity.

Every year the National Grange becomes more unlike any of the other farm groups. Roughly half of the Grange membership is nonfarmer, and its policy positions reflect many issues of social and farm reform from several rather socially conservative perspectives. The staff usually lobbies on member-inspired issues in the least obtrusive manner possible, usually by mail. Grange lobbyists, with their own personal interests focused on farm issues, occasionally surprise many observers by being active on Capitol Hill, working against something like

TABLE 2.2
PRODUCER/AGRIBUSINESS ORGANIZATIONS

PRIMARY	SUPPORT
National Broiler Council	Agriculture Council of America
National Cooperatives Business Association	Farmland Industries
National Cotton Council	National Association of Agricultural
National Council of Farmer Cooperatives	Employers
National Milk Producers' Federation	
National Rural Electric Cooperatives	
Rice Millers' Association	
Sunkist Growers	
United Fresh Fruit and Vegetable Association	

mandatory production controls or spearheading support for a farm bill provision that they find especially attractive but which is ignored by others.

The activist commodity groups are split by more than just product distinctions. The American Soybean Association and the National Cattlemen's Association staffs take great pride in their basic identification with free-market principles, pointing rather disdainfully to the Farm Bureau's more pragmatic support of government price programs. The Wool Growers and the Pork Producers might like to identify with such principles, but their lobbyists recognize that doing so invites a worsening of already threatening levels of imports from foreign producers. As a result, these relatively small offices concentrate on direct-benefit programs such as a checkoff to raise revenue for pork product promotions or the establishment of import quotas in trade bills.

Wheat, corn, peanut, and sugar groups are all primarily involved in protecting and enhancing provisions of rather different commodity-specific farm revenue programs. All are open advocates for government intervention in agricultural policy to the extent that many of their staff members are identified as among the preeminent progovernment leaders in the agricultural lobby. The Wheat Growers and the Corn Growers operate as integrated national organizations. The Peanut Growers Group, on the other hand, is a coalition of three long-standing regional groups. The four distinct regionally organized sugar producer groups, along with some smaller sugar organizations, operate similarly but on a less formal basis.

PRODUCER/AGRIBUSINESS ORGANIZATIONS

Producers are actively represented in agricultural policy through other organizations, especially cooperatives and associations with co-op memberships. This arrangement, while enhancing the representation of farmers, further complicates the meaning of farm interests by presenting a broader range of policy issues. The Rice Millers' Federation, National Milk Producers, and National Cotton Council are the major representatives of their respective commodity programs. The United Fresh Fruit and Vegetable Association attends to a wide range of industry policy problems from pesticides to trade and transportation. In each instance, programs of basic concern are more complex than those addressed by such groups as the National Corn Growers Association. When producers also process and even distribute, marketing orders may become a major component of commodity programs, as is the case for milk. When other middlemen such as shippers and brokers are included in the membership, as is the case with United Fresh Fruit, a basic but controversial commodity program such as navel orange marketing orders, may be consciously avoided because membership interests are more divided. In these instances, another organization such as Sunkist Growers may move to fill a representational void. Only when the industry is fairly well integrated by giant producers such as Holly Farms and Perdue in the broiler business can an organization generally represent producers and middlemen with ease. Even then, if oversupplies threaten the industry, internal adjustments will be hard to make and an organization may find its policy approach challenged by any dissident who is a very important member. This was indeed the case when Land O'Lakes broke ranks with the National Milk Producers and did not endorse the 1985 dairy program. In short, producer-middleman organizations need to be effectively brokered by their leadership in ways uncharacteristic of the producer groups.

To some extent, internal brokering between interests is also necessary for the other large national cooperative associations, as well as for Farmland Industries. While the National Cooperatives Business Association (formerly the Cooperative League of the USA), National Council of Farmer Cooperatives, and National Rural Electric Cooperatives primarily represent farm producer-supplier interests, these groups can also get enmeshed in middleman issues. The old Cooperative League was always a supportive dairy program player, as was the heavily staffed Rural Electric Cooperatives. The National Council, in addition to dairy supports, has shown broader interest in farm issues and as a result is often called another general farm group. For example, the council has been involved in promoting marketing orders and, due to similar benefits, advocated the concept of advance marketing loans in the 1985 farm bill.

The final two organizations having both producer and agribusiness members are quite different from the cooperatives or from one another. Nonetheless, because of possible membership tensions, these groups take very narrow and low-profile policy approaches. The National Association of Agricultural Employers (NAAE) represents businesses, trade associations, and farmers on such matters as the Farm Labor Act, pesticide regulation, and foreign workers. The NAAE has not attempted to play a leadership role on any of these issues, serving

TABLE 2.3

AGRIBUSINESS MIDDLEMEN ORGANIZATIONS

PRIMARY	SUPPORT
American Bakers Association	American Peanut Products
American Frozen Food Institute	Manufacturers, Inc.
American Meat Institute (AMI)	Chocolate Manufacturers' Association of
Corn Refiners Association	the U.S.A.
Food Marketing Institute (FMI)	National American Wholesale Grocers'
Grocery Manufacturers of America	Association
International Association of Ice Cream	National Frozen Food Association
Manufacturers/Milk Industry Foundation	National Independent Dairy-Foods
National Food Processors	Association
Tobacco Institute	National Soft Drink Association
	National Soybean Processors Association
	Peanut Butter and Nut Processors'
	Association

instead as a supportive coalition partner on what are divisive subjects within both the organization and the agricultural lobby. The Agriculture Council of America has also played a coalition role, primarily in raising issues of general agricultural concern rather than actively addressing them. Its concerns have shifted to finding ways of addressing export problems from a joint producer-agribusiness perspective. The small staffs of both organizations have been in a state of flux as they struggled with identification of their basic interests. The council's 1986 hiring of former Secretary of Agriculture Orville Freeman as president was an attempt to resolve some of those problems by enhancing the staff's brokerage abilities.

AGRIBUSINESS MIDDLEMAN ORGANIZATIONS

The most obvious thing about the middleman organizations is their place in the center of the food chain. Of seventeen trade associations, none primarily represents commodity merchants, only the National American Wholesale Grocers Association is a distributor group, and only the Food Marketing Institute (FMI) represents retail interests. Of the remaining fifteen organizations, nine are nominally processor associations, and six represent food manufacturers. Most of the processor associations, with the exception of the Corn Refiners' Association, represent the food manufacturing concerns of members as well. Advances in frozen foods have enhanced this integration. One organization, the Tobacco Institute (TI), integrates its commodity industry even further; it directly represents importers through wholesalers and exporters and claims to be the voice of tobacco growers and consumers as well. In that sense, TI is informally much like the National Cotton Council, with its encompassing organizational structure serving as an umbrella for an entire set of interrelated businesses.

The actual policy involvement of many of these trade associations is not nearly as broad as even their members' business concerns would suggest. The American Bakers Association is identified primarily with wheat issues. For years, the Corn Refiners were known only for their interest in corn sweeteners and, as a result, the sugar program. Agricultural policymakers in Washington almost instinctively link the International Association of Ice Cream Manufacturers and its sister organization the Milk Industry Foundation, the Chocolate Manufacturers' Association of the USA, the National Soft Drink Association, the National Independent Dairy-Foods Association, FMI, the Grocery Manufacturers of America, the Wholesale Grocers, the Peanut Butter and Nut Processors' Association, and the American Peanut Products Manufacturers to what they perceive to be an orchestrated opposition to dairy and sugar programs that incidentally—in candy making—involve peanuts. In practice, however, these organizations are active in other issues no less important to their members' businesses. Their export, regulatory, product purchase, and food safety lobbying fails to receive the widespread attention given to controversies over commodity programs. These other issues take up almost all the agricultural policy time and attention of the American Meat Institute, National Food Processors, American Frozen Food Institute, National Frozen Food Association, and National Soybean Processors Association.

SUPPLIER/FACILITATOR ORGANIZATIONS

While important common interests are identified with many middleman organizations, supplier and facilitator organizations tend to take their own approaches to public policy. Three highly visible facilitators include the Farm Foundation, the National Center for Food and Agricultural Policy of Resources for the Future, and the American Enterprise Institute. None are, in any traditional sense, lobbies with specific membership agendas; each is an experienced research and educational foundation involved with sponsoring agricultural policy research and holding public forums to disseminate research findings. Their reputations for expertise are based on open and candid discussions of topical agricultural issues. A wide variety of agricultural specialists, especially economists, participate. As a result, conferences and publications are followed closely by a variety of other agricultural policy participants. This attention, combined with each organization's desire to produce highly credible research and encourage the emergence of agricultural leaders, has given the three organizations all the information-based influence of any private interest.

Though these foundations tend to view their interests similarly and share a free-market approach to agriculture, other closely related organizations from this category are quite different. The Independent Bankers Association of Amer-

TABLE 2.4
SUPPLIER/FACILITATOR ORGANIZATIONS

PRIMARY	SUPPORT
Farm Credit Council	National Association of Colleges and
Farm Foundation	Land Grant Institutions
Fertilizer Institute	American Bankers Association
Independent Bankers' Association	American Enterprise Institute
National Agricultural Chemicals	American Seed Trade Association
Association of America	Chicago Board of Trade
Resources for the Future (National Center	
for Food and Agricultural Policy)	
United Farm Workers	

ica and the American Bankers Association represent important agricultural lenders and operate specialized internal units to monitor agricultural conditions. The Independent Bankers Association, which has directed more of its attention to borrowers' credit problems, argued for production control legislation in 1985. The American Bankers Association, with its membership of larger banks having outstanding foreign investments in agriculturally developing nations, has been more expressly concerned with foreign policy aspects of food and fiber politics. The third financial organization, the Farm Credit Council, with its membership from the cooperatively owned Farm Credit Banking community, emerged under the auspices of the National Council of Farmer Cooperatives and moved almost immediately to obtain federal assistance for these lending institutions.

The supplier associations—the Fertilizer Institute (FI), National Agricultural Chemicals Association, and American Seed Trade Association—have also reacted differently to farm problems of the 1980s. The chemical and seed associations have long been active on such product issues as the Federal Seed Act and the Plant Variety Protection Act. However, the Fertilizer Institute, once a foremost opponent of the Soil Bank Program, generally avoided agricultural policy issues during the 1960s and 1970s until its members felt the effects of the 1983 Payment-in-Kind (PIK) Program with its restrictions on planting and, as a result, fertilizer usage. During the 1985 farm bill proceedings, the FI worked to develop an image as the ranking agribusiness opponent of federal commodity support programs.

The other supplier/facilitator organizations gained their reputations in equally diverse ways. For many years the Chicago Board of Trade has quietly supported FI's current principles. The American Association of Colleges and Land Grant Institutions gets most involved as a broker in mediating such research and product controversies as the one over the Federal Insecticide, Fungicide and Rodenticide Act (FIFRA). The United Farm Workers is a more activist organization than either of these but does most of its policy work through direct pressure on farmer growers. It relies on ideological sympathizers to pick up on its well-publicized grassroots work to promote federal policy reform on its members' behalf.

PUBLIC INTEREST GROUPS/RELATED ORGANIZATIONS

In a sense, each of these groups is acting as a public interest group (PIG) on agricultural issues since none of them is directly pursuing its own members' material interests by advocating policy reform. These groups are all private interests with their own unique visions of what they want from American agriculture. The United Auto Workers (UAW) and the National Council of Churches, the latter through many different outlets, have been active in articulating the problems of the farm crisis in grassroots support of farm protest groups and, to a lesser extent, in Washington. The National Catholic Rural Life Conference is similarly active in the Midwest and in state politics. UAW locals in farm equipment industries and rural churches, both of which have been grievously damaged by the farm economy, are intended to benefit from this activism, however, so all three organizations fall somewhat short of being true public interests.

Most of the other organizations are prominent leaders in specialized issue areas related to agricultural policy. American Farmland Trust emphasizes conservation issues. The Environmental Policy Institute, alone among the ten national environmental groups, is concerned with the effects of agriculture on the environment. Public Voice is the consumer advocate for food safety and pricing issues. The Food Research and Action Center emphasizes attention to food stamp programs but formerly was very active on child nutrition matters. Bread for the World engages in the same advocacy on behalf of world hunger programs, especially Food for Peace. Livestock and laboratory animal conditions are increasingly the concern of the Humane Society of America after years of greater visibility by the Animal Welfare Institute. Each of these six organizations benefits politically from its specialized subject knowledge and the reciprocal support of several other groups involved in related matters but less active in agricultural issues. When issues of interest are identified, these PIGs are usually able to mobilize several other closely allied groups as well as grassroots activists.

Two other public interest organizations try to work in similar ways. The Capital Legal Foundation is attempting to exercise comparable issue leadership on marketing order policy but has yet to make the issue politically prominent. The Farmworker Defense Fund (FDF), through Migrant Legal Action, supports farm worker activism. However, the FDF finds policymakers themselves exercising most of the Washington-based leadership through the occasional prodding of the United Farm Workers.

Only one of these groups, Congress Watch, operates across issue areas and with a reputation for generalized expertise. This Ralph Nader affiliate attends to

TABLE 2.5
PUBLIC INTEREST GROUPS AND RELATED ORGANIZATIONS

PRIMARY	SUPPORT
American Farmland Trust	Humane Society of America
Environmental Policy Institute	Bread for the World
Food Research and Action Center	Capital Legal Foundation
National Catholic Rural Life Conference	Congress Watch
Public Voice	Migrant Legal Action Program
	(Farmworker Defense Fund)
	National Council of Churches
	Rural Coalition
	United Auto Workers (UAW)

various consumer food issues. The reputation of Congress Watch is more a product of its Nader ties and alliances than of any highly specialized expertise.

AGRIBUSINESS AND FOOD FIRMS

Given the hundreds of firms involved with agriculture, it is surprising that few have a reputation for political activism beyond their trade association memberships. For the most part, these firms are seen as politically reluctant. Many are not very different from the five listed here in the "support" category. ConAgra sometimes advances an innovative trade proposal. Pioneer works in a leadership capacity on matters of importance to the American Seed Trade Association. Monsanto does the same through the National Agricultural Chemicals Association. John Deere traditionally argues for moderation of farm bills. Ralston-Purina is often involved with several other organizations from outside agriculture to articulate an agribusiness position on a taxation or trade issue.

Five other firms, however, are different from the others in that they have reputations for investing both more thought and more money in public-policy aspects of corporate public affairs. Two diversified merchant-processors have reputations for being among the most active and involved individual business firms in Washington. Cargill has been such a company for years, whereas Archer-Daniels-Midland (ADM) is a relative newcomer.[29] Cargill favors a general freedom in its grain trading activities, but ADM openly supports governmental intervention if that suits the firm's needs. Safeway Stores, with its Washington headquarters, has also been a long-term activist organization on agricultural issues. Safeway's policy concerns have been somewhat broader than those of Mars and Pizza Hut, two other firms that have recently sought to maximize profits through political activism. Both of the latter firms are seen by

TABLE 2.6
AGRIBUSINESS AND FOOD FIRMS

PRIMARY	SUPPORT
Archer-Daniels-Midland	ConAgra
Cargill	John Deere
Mars	Monsanto
Pizza Hut	Pioneer
Safeway Stores	Ralston-Purina

policymakers as corporate leaders of the coalitions opposing dairy and sugar pro-grams. What makes them and the other three firms unique is not their position taking, however. Their uniqueness, in the eyes of policymakers, is due to their representatives' willingness to explain, defend, and provide a sound and specific rationale for the positions they advance.

POLICY AND INTERESTS: SOME CONCLUSIONS

During the Washington phase of the interviews, an early trend developed in re-spondent questions. Lobbyists and policymakers alike noted the absence of groups who had apparently left the scene. Where, the respondents asked, were the Grain Sorghum Producers Association, Rural America, the Midcontinent Farmers Association, cattlemen from Texas and the King Ranch, big-city mayors, the poor people's lobby, and several other organized interests that were not very long ago considered integral parts of the agricultural lobby? In each instance, phone calls or personal visits indicated that the group was still in existence and at least somewhat involved with its patrons or members but no longer involved in any ongoing manner in agricultural policy. For some, events and circumstances had changed their need to actively lobby. Others lacked the organizational capac-ity to attempt it. Both the questions and discoveries underscored the temporal nature of the agricultural lobby.

This lobby grew and changed, and it keeps changing in response to a large and aggressive environment. While all organizations might have at least latent political agendas resulting from their members' shared interests, not all of these agendas became active at once. The evidence in this chapter suggests three rea-sons for this phenomenon. First, not all political agendas are as important as the other membership activities that occupy leaders, staff, and organizational re-sources. Second, not all political agendas are relevant to the issues being dis-cussed. Third, there may be no chance of certain items' winning.

One example will suffice until these points are explained further in later chapters. At the height of the controversy over the 1985 farm bill, a reduced

Rural America staff was working to maintain that organization by contracting to do technical work on housing and transportation problems for rural local governments. Two farm bills ago, eight years earlier, Rural America was investing all of its scarce resources in promoting a concern for the social and economic conditions of nonfarm rural society. While momentarily startling, it was not surprising that Rural America's staff was not following or even much aware of the current status of the farm bill in 1985. The staff had to worry about providing the traditional kinds of services that had kept this primarily non-member-funded organization together. Second, had it tried to influence the 1985 farm bill, Rural America would have found that declining exports and the economic problems of farmers had at least temporarily detracted from government concern about general rural problems. That is to say, the 1977 and 1985 farm bills existed in two entirely different policy contexts. Conditions in 1985 made it harder for Rural America both to raise money and to gain political attention. As a result, while it may well have been possible to ask serious questions about the viable future of rural communities in a depressed farm economy, prospects for corrective government action were bleak. No reason existed, at that time, for Rural America to waste its limited staff time, attention, money, and other political resources. It would only have lost anyway.

Other agriculturally interested organizations reflect the same moderation in developing their political demands and responses. So, while the food and fiber systems are extensively organized, the components are not equally represented in agricultural policy, nor are all possible issues raised by even the most active lobbies.

Lobbyists for both active and relatively uninvolved agricultural interests believe overwhelmingly that policymakers perceive most issues as the domain of farm organizations and accord them special legitimacy when farm conditions are affected. Under such conditions, farm organizations are most likely to organize around agricultural issues, since they are most likely to satisfy their members and win by so doing. Even the farm organizations exhibit moderation in what they choose to represent. Rather than competing to represent the same issues and policies, established organizations and most new groups occupy distinct policy niches. As noted earlier, this tendency is as old as the agricultural lobby itself.

Agribusiness, whether through trade associations or single firms, selects more carefully which issues to support or challenge. Two characteristics are evident. First, single organizations generally tend to occupy much broader policy niches, so almost all facets of the food system and related issues are provided some political coverage. They avoid representational duplication. Second, where political attention overlaps, it does so on a coalition basis centered around what are seen as vulnerable and costly public policies especially open to change. In the 1980s, sugar and dairy programs elicited this response. In the 1950s, support programs—and later the Soil Bank—were targeted. In the same way, in the 1920s, businesses organized against laissez-faire farm economics.

Even greater specialization and tighter coalition support characterize the newest organizations active in agricultural issues, the public interest groups. While these organizations have expanded rapidly by cultivating the patronage of established organizations, foundations, and philanthropists, there is little evidence that they create great policy confusion through their expansion into the agricultural policy arena. On the contrary, the public interest groups operate as reformers within their own highly specialized policy niches. As a result, these groups bring a rational, orderly, and efficient approach to their lobbying.

Taken together, agricultural interests are highly diverse and capable of producing both policy conflict and mixed political message. The agricultural lobby's collective problems make its overall political involvement rather self-limiting, however. As a result, while the agricultural policy process has become highly complex because of the existence of so many private interests, there is an underlying order that many if not most interests work very hard to preserve. The next chapter turns to the behavior and organization of agricultural interests as they attempt to exist within that universe of interests.

3

The Policy Process
and Interest Representation

More than any other policy area of American government, agriculture has been portrayed as closed from the political intervention of "outsiders." The identity of agricultural "insiders" has changed. From the Farm Bloc era and into the 1950s, insiders included a small cadre of farm state legislators, ranking USDA officials, and representatives of the Farm Bureau, the NFU, and the Grange. At times, and on such major issues as the Brannan Plan, widespread partisan conflict between the Republican-oriented Farm Bureau and the Democratic-allied Farmers Union broke out.[1] When such disagreements could not be resolved by the insiders themselves, cues emanating from within the agricultural establishment—such as the advice of large-scale producers to friendly senators—nonetheless structured final policy decisions. Policymakers listened to, acknowledged, and backed agriculturalists even when the specialists were divided among themselves.

From the mid-50s to the mid-70s circumstances changed greatly, but control of the agricultural agenda remained almost unchanged. As noted, commodity groups, with their specialized knowledge and ability to build encompassing product-by-product coalitions on major legislation, gained representation and influence. Along with the congressional agriculture committees and commodity specialists within the USDA, they were the new insiders.[2] Instead of ruling on the basis of policymakers' electoral needs to satisfy a farm population that was no longer larger than any other, the new insiders controlled information and increasingly governed on the basis of their issue expertise. By then, power in Congress had been fragmented into committees and subcommittees, so congressional dependence on these decentralized units soon followed. Throughout Congress committee decisions were ratified more routinely on the basis of trust and reciprocity between legislators than as an outcome of floor debates.[3]

The expansion of bureaucracy, along with the continuing escalation of policy analysis within agencies, was integral to this shift.[4] Career civil servants shared policy leadership.[5] Along with congressional experts and lobbies whose specialized interests focused intensively on policy decisions, they came to be seen as the three principal components of highly specific subgovernments. These subgovernments existed in a political universe in which both the technical complexity and the sheer number of issues to be resolved forced decentralization in policymaking.[6]

For many years, central authorities in the administration and Congress were unable to control agricultural policy. The White House, the Bureau of the Budget, House and Senate leaders, and partisan officials were the outsiders for two distinct generations of agricultural politics. Legislators from other committees, nonfarm agencies, and sometimes even secretaries of agriculture occupied the same outsider status because of a lack of access to the seat of agricultural policy deliberations.

Conditions have once again shifted, however. The complexity of agricultural policy—with its farm, food, and foreign emphases—expands decisions of interest beyond a single committee and its subcommittees in each house. In addition, a spate of provisions with additional program benefits has been added to major farm bills since at least 1973 in order to gain sufficient legislative and constituent interest to pass that legislation. Mutual support of legislative colleagues has not been enough.[7] Budgetary reforms, as well as concern about expanding budget deficits, have vested important decisions in a revitalized Office of Management and Budget and an equally active Congressional Budget Office. For these and various other reasons, the White House is more likely to be directly involved. Where there was once a single locus of power, multiple centers of influence now exist, with the authority at least to exercise important checks on the agricultural establishment. In short, distinctions between agricultural insiders and outsiders have lost much of their policy relevance.

LOBBYISTS AND THE POLICY PROCESS

The shift from closed patterns of agricultural policymaking to a contemporary situation in which outsiders are included can be understood as an expansion of both relevant issues and newly important participants. Alternatively, this change can be seen from the perspective of those lobbyists who, as only a few among several sets of policy participants, are attempting to influence the process. In many ways, the lobbyist's perspective is more meaningful than a conceptual explanation of an evolving policy process; participants are unhindered by a need to develop a theoretical explanation of who does what. Lobbyists are also more specific in explaining what they see than are nonparticipant observers who want to generalize about behavior and influence. The lobbyist simply asks, usually on a case-by-case basis, "Whom shall I contact and why?"

The need for data on lobbyist opinions can best be seen by looking at policy explanations formulated by those who have studied agriculture. Subgovernment theory, as noted above, has been of central importance in explaining agriculture. Theorists such as Theodore Lowi have relied on it, as have critics of the policy process such as Charles M. Hardin and Clifford M. Hardin. Lowi developed his attack on liberal government on the basis of its inability to deal with broad issues of national consequence.[8] Because of their self-serving interest in distributing program benefits among policymaking partners, narrowly specialized policymakers ignore such issues, Lowi argued. Hardin and Hardin, both especially perplexed by Congress's continued commitment to price support programs, blamed subcommittee government with its bureaucratic and interest group linkages.[9]

Explanations of this sort appeal to problem solvers who confront agricultural policy because they can easily be generalized. In application, the Lowi-Hardin-style critique can be proposed as a model. Neil Meyer and William Dishman did precisely that by developing a cluster-based diagram of specialized agricultural policymaking and suggesting that it be used as a road map for those who want to become involved in political decisions.[10] James T. Bonnen used the emergence of subgovernmental relationships and other decentralizing factors to decry the demise of the political party and with it the mediating impact of partisanship in stabilizing policy positions.[11]

To the extent that each made important contributions to policy awareness, these observers are not necessarily wrong in their conclusions. Agricultural policy, in a historical sense, has been a perfect new example of decentralized policymaking, tripartite political relationships, and narrow, farm-directed policy. Agricultural policymakers appear to have defied centralized control, spurned cutbacks in ineffective programs, continued to promote programs to minimize farm risks, and ignored many important food problems. While correct in identifying important policy implications, these and other observers present an incomplete picture of the dynamics of the agricultural policy process. In making policy decisions, behavioral nuances are as vital as trends, and they cannot be found by tracing back from a macrolevel view of policy outcomes. The policy context is no less important, since factors such as the economy and environmental hazards are just as likely to bring issues to the political agenda as are interactions between interest groups and other political spokespersons. The subgovernmental interpretation of agriculture emphasizes a narrow set of participants. Diverse analyses of agricultural policy have been too caught up in theoretical paradigms and explanations to observe the process closely. On the basis of limited information, researchers presume a great deal about the adaptation of policy insiders to the wave of 1970s outsiders in agriculture. The policy process and its changing dynamics seem to have produced far too many checks, balances, points of access, hurdles, and roadblocks for anyone to generalize much about an agricultural lobby before sharing its perspective.

During interviews for this project, lobbyists representing organizations from the food and fiber system discussed the people with whom they worked on policy decisions in the 1980s. Their responses revealed much about the transitional phase of agricultural policymaking that began, for all practical purposes, with the 1973 farm bill. Although the agricultural establishment plays a dominant role in issues of primary interest to most of these organizations, many relevant agricultural programs and policies are negotiated elsewhere. In addition, new strategies have evolved as a result of beliefs about what benefits previous agricultural outsiders can bring to most of these organizations. One target is the single legislator, who, regardless of position, works actively on behalf of constituents. A second target is the executive office and the influence that a sympathetic White House statement can have in creating a mood of policy support. Additional expertise is also a target. Lobbyists search diligently for supplemental evidence in support of group positions both inside and outside agriculture. Taken together, these factors make for a more comprehensive approach to agricultural policy than might a policy evolution away from a monolithic agricultural establishment. The changes of the past decade have been more a revolution than an evolution of representation. Politics in the 1980s is reminiscent of the pluralism and diversity of purpose that characterized the four-group agricultural lobby during its early Washington years in the 1920s.

COMMITTEES IN CONGRESS

Congress remains the lobbying center, even though most lobbying-staff growth during the 1970s was in positions designated for agency lobbying. According to respondents, litigation as a proactive lobbying tactic gains more proponents each year. Nevertheless, congressional contacts are the most frequent for all but a small number of groups. Almost none of these organizations ignore the agricultural committees, especially in the House of Representatives, where coalition building is seen as essential to producing legislation. As critics suggest, the Senate Subcommittee on Agricultural Production, Marketing, and Stabilization of Prices and the four House commodity subcommittees receive most of this attention since they make recommendations on price and loan support programs. Commodity groups whose members benefit from specific programs acknowledge that their lobbyists do most of their work with subcommittee members. Agribusiness interests concerned with specific commodity programs as well as organizations that generally favor a more free-market agriculture usually attempt to work first with these same subcommittees. Organizations that want to promote alternative omnibus farm bill proposals noted that the optimal strategy is to do so through the subcommittees that handle commodities. The remaining Senate and House subcommittees have similar interest group constituencies on such recurring issues as credit, nutrition, forestry, marketing, foreign agriculture, and agricultural research.

Other committees and subcommittees are no less important than the ones for agriculture. Appropriations committees, with their agricultural subcommittees, are of vital concern when programs must be funded, since some financial limitations will affect the degree to which services and benefits can be provided. In this instance, lobbyists count on close cooperation between agriculture and appropriations committee members rather than on a great deal of direct influence from interest groups. Many organizations favor targeting the appropriations committees as an obstructionist tactic, however. Federal budget deficit concerns have enhanced the use of appropriations as a potential lobbying target, increased conflict between agricultural interests, and raised levels of organizational anxiety over attaining necessary funding levels.

Jurisdictional overlaps between committees and subcommittees have complicated lobbying for many organizations, including traditional farm groups. So has the proliferation of issues of concern to those within the agricultural lobby. Some farm problems may be dealt with by nonagricultural subcommittees with responsibilities for toxic substances, small business and family farms, labor, banking, trade, consumers, energy, agricultural taxation, estates, international regions, health and safety, public lands, water resources, and economic development. Environmentalists, consumers, and similar public interest groups that feel they get a less than favorable reception in agricultural subcommittees often work hard to direct issues to those legislative centers. General farm and some commodity group lobbyists also approach these legislative units as alternatives to the agriculture committees.

Interests with special concerns of foreign debt, trade policy, and the stability of specific nations have the most formidable reputations for taking agricultural issues to committees that deal with foreign affairs and economic policy, though. Banking interests, multinational business firms, and the U.S. State Department are cited frequently as interests that intervene in agricultural issues while attempting to circumvent agricultural subcommittees.

This strategic attention to nonagricultural committees is less common than the ongoing need of many farm and food interests to spend time on other matters.[12] Most agribusiness interests, for example, work more with legislators from the Senate Finance Committee; House Ways and Means Committee; Senate Commerce, Science, and Transportation Committee; and House Energy and Commerce Committee than with agricultural committee members. Since such issues as estates, taxes, and fuel are basic to agricultural production, more farm group lobbyists are being hired to assume the same responsibilities.

WORKING WITH LEGISLATORS

Research stresses the importance of congressional committees and subcommittees and the accompanying decline in the influence of central legislative leaders. Chairpersons, the literature suggests, are most likely to be the ones who make the policy process work, but they are not alone as political catalysts within the committee structure.[13] John Kingdon and others suggest that rank-and-file legislators also serve as opinion molders and, in effect, de facto policy leaders for their colleagues in distinct issue areas.[14] This means that a fairly large number of legislators can effectively direct portions of what becomes law, even when their individual proposals have little relationship to a coherent agricultural policy. This raises a yet unanswered question, "How do lobbyists attempt to make the committee structure work to their advantage?"

Ideally lobbyists want to find a subcommittee member willing to become involved with the issue of interest. Chairpersons usually are not targeted more than anyone else. Though they may be in a position to exercise the greatest leadership and control over their own subcommittee agendas, they have busier schedules than their colleagues on matters of policy relevance. As a result, lobbyists tend to concentrate on finding any member recognized as diligent in his or her subcommittee work. Then they look for a legislator who will listen to the organization, has time to spend on the issue, and is likely to be given attention by congressional colleagues. Finding a legislative spokesperson is a highly selective process and one that is viewed by organizational representatives as especially important in promoting a legislative and frequently an administrative proposal. Most organizations find that their choices must often fall short of the ideal, since there are far more proposals affecting agriculture around Washington than there are active and respected legislators.

In view of this shortage of useful legislators, organizations frequently proceed slowly on proposals that their staff or members might like to advance. Sometimes this means ponderously promoting an idea until it becomes a recognizable policy alternative. At other times lack of an acceptable legislative spokesperson effectively kills a proposal inside the organization, even though it may still be thought of as a very desirable public policy.

When the lobbying goal entails defeating a proposal or reforming a current program, strategies for working with members often change. Soliciting one or more spokespersons remains the preferred method. Since less time and fewer resources are necessary for building rather then overcoming legislative hurdles, there is less concern about the legislator's capacity for hard work and personal attentiveness when an issue needs to be defeated. If an active congressional ally cannot be found, however, continued opposition remains a feasible option. Organizational representatives need only to collect nay votes and gather opposition rather than overcome the obstacle of bringing legislators into a proactive campaign. This is how lobbyists who depend on legislative followings conceive of a defensive advantage. Lobbyists feel that their own political goals are increasingly disadvantaged as the cost of a legislator's support goes up in terms of the personal time and attention that he or she needs to commit to an issue. One veteran lobbyist summarized the situation: "Congressmen like to be able to agree with you by saying no."

Given the emphasis that lobbyists place on the committee structure, are members who do not occupy committee or subcommittee seats sought as opinion leaders? There was a surprising concurrence that they can be. Only two of the respondents who addressed this subject dismissed the option of going to a committee member with different subcommittee responsibilities or someone otherwise uninvolved in agricultural issues. Even these unenthusiastic respondents agreed that other organizations used such legislators effectively, however. The development of sophisticated public relations techniques is credited with enhancing this strategy.

If carefully coordinated and able to strike a responsive chord, public relations campaigns can bring an awareness of an issue or policy option directly to the public and to media-conscious policymakers by forcefully dramatizing it. Conferences, demonstrations, protests, or threatened organizational action may attract media coverage. Lobbyists with access to reporters or editors may simply get the media to report on information they provide. Paid advertising, in which both the importance of the problem and the social implications of neglect are reported precisely as the group wishes, provides additional opportunities and advantages. After publicity groundwork has been laid by a lobby and widespread attention is focused on the problem, almost any credible legislator can become the spokesperson for the issue by alluding to widespread constituent impact. Indeed, as lobbyists see it, a well-orchestrated media campaign with well-distributed print and electronic coverage elicits a variety of legislators happy to jump on an issue bandwagon. Under such circumstances, subcommittees find it difficult not to respond by at least assisting in formulating a policy response.

Lobbyists also believe, for targeted subcommittee members as well as rank-and-file legislators, that constituency ties are generally of greater consequence than familiarity or expertise with the issue. The two may be simultaneous factors, though. Farm state members are targeted by farm groups. Commodity organizations do not often look for legislative spokespersons from states in which those products are not grown in abundance. Both trade associations and firms feel most comfortable activating legislators from districts or states in which an industry is well established. During the political maneuverings on the 1985 farm bill, for example, Representative Dan Glickman actively opposed dairy programs and, as a result, was frequently referred to as the "Congressman from Pizza Hut." Pizza Hut's corporate offices were in Glickman's Kansas district, and that firm had been instrumental in organizing opposition to dairy programs. Similarly, Archer-Daniels-Midland has been able to cultivate an unusually close working relationship with the Senate leadership in the mid-1980s, in part because Majority

Leader Robert Dole comes from Kansas, a state that boasts ADM corporate offices. The availability of strategies that direct policy attention beyond the subcommittee win support from most lobbyists because, though costly, these tactics provide options when a subcommittee appears firmly opposed to a group's position.

AGENCY LOBBYING

There is much in common between lobbying Congress and the administration, despite the bureaucracy's lack of need for electoral assistance. First, many contacts are made in both branches on the basis of expertise. While much attention to Congress focuses on the subcommittees, interests seek out experts who play similarly specialized roles within agencies and bureaus. The agency list is extensive. For instance, commodity programs are housed in the Agricultural Stabilization and Conservation Service (ASCS), and much discretion is vested in the ASCS. Commodity experts with important advisory and analytical responsibilities that USDA officials and Congress rely on extensively are housed in the Economic Research Service. Some commodity lobbyists consider the ASCS and ERS offices extensions of their own workplaces. None of the commodity organizations ignore these bureaucrats. Other specifically targeted agencies abound, since nearly every USDA unit has some regulatory authority over some agricultural programs, either authorizing benefits or restricting some type of resource use. The Foreign Agricultural Service (FAS), for instance, provides trade assistance programs; and those groups that are organized to enhance commodity exports cultivate FAS ties. The Agricultural Research Service gets the same comprehensive attention from groups and associations that have concerns affected by research and technical questions.

A second feature common to congressional and administrative lobbying has to do with the intent of those making the contacts. Lobbyists want bureaucratic advocates. They also wish to neutralize any perceived opponents. To achieve both goals, agricultural lobbyists engage in the same search for administrative spokespersons as they do in Congress. Their principal stock in trade, despite and because of the bureaucracy's research orientation, is supplemental information and assistance in making programs work. Organizations such as the Rice Millers' Association demonstrate consistency in mastering the type of economic analysis that policymakers expect of economists in the ERS. Bureaucrats are grateful to receive information in support of their own research capacity.

Legislator intervention is the third feature that administrative and congressional lobbying share. Congressional members, with their control of both programs and budgetary purse strings, can help mold opinions as easily in administrative agencies as in the legislature. Hence lobbyists without administrative contacts or those who cannot convince bureaucrats to consider certain policy options often go to legislators for a bureaucratic response. Although

USDA agencies have limited regulatory responsibilities, legislators can also be used to intervene in promoting or challenging specific regulations.

The increased regulatory involvement of government in food issues perhaps does more than anything else to differentiate administrative from congressional lobbying. While almost all agricultural lobbies find themselves with some need to deal with House and Senate agricultural committee members, nearly half of the respondent organizations from Tables 2.1–2.6 do not lobby the USDA. Respondents from almost all of the organizations mentioned a need for some bureaucratic lobbying outside the Department of Agriculture. In other words, administrative lobbying is much less focused on a common set of institutional actors with traditional agricultural ties than is congressional lobbying.

Regulatory growth, in terms of imposing restrictions on agricultural production and products, has occurred principally in environmental issues, food safety, and occupational safety. As a result, agricultural lobbies from throughout the food system have had to interact with the Environmental Protection Agency, the Food and Drug Administration, and the Occupational Safety and Health Administration on a wide range of issues.

With the expansion of agriculture into foreign trade and export issues, agricultural interests have felt it necessary to expand agency lobbying into new programs. The Farm Bureau, commodity organizations, trade associations, and trade firms all spend increased time in the Departments of Commerce, Transportation, and Treasury, in addition to the overlapping jurisdictions that have long involved farm groups with agencies from the Department of the Interior. Another administrative organization that many agricultural interests would like to lobby but which few approach directly is the State Department, which is responsible for many economic and trade policies that lobbyists believe would be decided differently in different agencies. The general opinion of embittered agricultural representatives is "You can't lobby State." Although it may be difficult, some of the more creative organizations such as Archer-Daniels-Midland still give it their best effort.

LOBBYING NATIONAL LEADERS

Studies of leadership in national government, of which there are many, indicate that White House and congressional leaders can and do impose their individual wills on specific programs if officials give them a sufficiently high priority. Political observers who concentrate on the restricted powers of national leaders or the proliferation and decentralization of issue management may neglect the fact that on some policies a president or house majority leader personally makes an unusual effort to structure a policy outcome. Kay Lehman Schlozman and John T. Tierney found that, especially for the White House, lobbyists have a renewed appreciation for this potential.[15] Only 12 percent of their lobbyist sample believed that the White House, as a target, was unimportant.[16] The president, represented by his key advisers, was seen as very important for 55 percent of their respondents.

The agricultural lobby was only somewhat in agreement. Less than a quarter of the most active agricultural organizations targeted the White House in the 1980s. Why was there a difference? First, farm programs and closely related issues are still seen as institutionalized in Congress. Most provisions are also considered too minor to worry the White House, so Washington-based farm organizations, with few exceptions, see little reason to lobby presidential assistants. Second, the Reagan administration has been seen as specifically hostile to farm programs. Even conservative groups alluded to a "Reagan-climate" of hostile opposition to farm supports. Only organizations such as the Farm Bureau and some agrarian protest groups thought it remotely worthwhile to lobby the White House and attempt to soften this opposition. Even these organizations shared the general consensus of "leaving Reagan to Congress." Third, the Reagan White House was seen as no more friendly to any of the public interest lobbies, with the possible exception of Public Voice, than it was to their frequent adversaries, the farm interests. As a result, the PIGs worked the Congress, lower agency levels, and sometimes the courts. Fourth, the White House was perceived as difficult to lobby on farm and food issues, since not even USDA Secretary John Block was thought to have access there during most of the Reagan years.

Who from the agricultural lobby targets the Reagan White House? And why? Most organizations have been trade associations and agribusiness firms. The Fertilizer Institute, for example, aspires to lead an industry-based coalition to eliminate commodity price supports. Its free-market goals conform to those of the Reagan administration, and FI feels a need to associate itself closely with White House goals in order to break what its lobbyists see as a "congressional stranglehold" on farm production issues. Some agribusiness lobbies see things the same way. Several firms, not all of which share policy views, are active at the White House on specific export and trade issues. Both Archer-Daniels-Midland and Cargill executives, for example, feel comfortable at the Reagan White House even though their policy preferences and lobbying tactics are often quite different. The shared attitude exists, however, that business interests with diverse perspectives on world conditions get a fair hearing at the White House. Business also goes to the White House because its lobbyists, like everyone else, find it hard to lobby the State Department without administrative ties.

Agribusiness by no means monopolizes the White House on agricultural issues, however. Different types of interests have brought a wide range of issues there. The Farm Credit Council considered White House agreement essential to the eventual passage of the 1985 farm credit bill. Foreign sugar interests sought administration support on increasing sugar import quotas by playing on the president's Caribbean Basin Initiative. Sunkist Growers and other cooperatives worked hard to neutralize possible White House opposition to marketing order legislation. In each of these instances, White House support was considered critical to these policy proponents.

Congressional leaders more frequently were targeted by a wide range of agricultural interests than was the White House, but in a rather different way. While these individuals were generally perceived as busy, several were seen as especially attentive to agriculture because of their own committee assignments. The Senate majority leader, Senate foreign relations chair, House majority whip, and House Democratic congressional campaign committee chair, as of 1985–86, sat on the agriculture committees. In addition, the House majority leader and House appropriations committee chair once played active agricultural committee roles and still felt comfortable stepping into committee matters. The House minority leader was also from an agricultural district and, as a result, was a natural target of farm and farm equipment interests.

The links between legislative leaders and agriculture make them a normal part of congressional lobbying. Because of the special conditions of the issue area itself and the policy beliefs of the president, the White House is less a target of agricultural lobbies than it is of nonfarm and nonfood interests. Issues that have involved the White House in the 1980s have tended to be either major agricultural ones that lose or more selectively narrow ones, somewhat beyond the controversial mainstream of farm and food politics. These issues are thought by lobbyists to need the extraordinary attention of the president in order to have any chance of victory. Lobbying the leadership, it must be concluded, is important for agricultural interests in the total dynamics of policymaking.

MARSHALLING SUPPLEMENTAL SUPPORT

Interest representatives frequently mentioned an upside to lobbying that cannot be divorced from its ever present downside. The upside is that some degree of support can usually be found among targeted policymakers. The downside comes because, once activated, policymakers are seldom under a lobbyist's control. With multiple-interest positions and a variety of nonlobbyist pressures with which to contend, targeted spokespersons may take a proposal and move in any direction. Supporters may reverse themselves in their voting commitments. Unpredictability, as seen by several lobbyists, results from the basic honesty and electoral concerns of most policymakers. Though no vote or support may be for sale, lobbyists openly joke about policymakers who claim to be for rent. The rentals are the only policymakers who are guaranteed to stand in place very long on a single issue or policy question.

To overcome uncertainty, lobbyists negotiate with other private interests at least as much as they do with policymakers. Coalition building, which is the subject of Chapter 9, constitutes a big part of this effort as lobbyists share information and contacts, work toward mutually agreeable positions, and jointly plan strategy.

Many other interactions between organizations could be summarized as yet another form of interpersonal rental agreement. While sometimes viewed negatively by both lobbyists and policymakers, these agreements are legitimate. The contractor is usually a traditional agricultural lobby, and the other party represents one of three forms of consultants. Some consultants provide direct lobbying assistance—and often litigation—on the basis of either issue expertise or political contacts in appropriate policy circles. Other consultants provide public relations skills to direct attention toward the problems or priorities of an interest group. The third type of consulting involves economic analysis, usually of a more technical or specialized kind than the organization's own employees are able or have time to produce. Most of this work is done by consulting firms that operate specialized units for agricultural work. Several of the firms are concerned only with farm, food, and fiber issues. Some provide all three services: lobbying, public relations, and analysis.

Consultants provide an important dimension to agricultural lobbying. If they gain policymaking access, organizations can purchase skills and experience that enable them to deal with issues and policymakers that they would otherwise neglect. Several lobbyists also see the use of consultants as a chance to monitor more constantly legislative and administrative spokespersons whom they might otherwise have to ignore and perhaps lose. Over three-quarters of the eighty-four most active agricultural lobbies hired one or more consultants for policy projects in the 1980s.

THE SUBSTANCE AND STYLE OF LOBBYING

The selectivity of organized interests in choosing issues on which to lobby reveals much about their usual behavior. Membership concerns that result in the use of organizational resources for personal services create policy niches for most agricultural interests. Things are not so simple for interest group leaders that they expend whatever resources they possess in pursuit of whatever public-policy goals they want. Though membership problems limit both the designation of political priorities and the mobilization of resources, other factors have similar effects. It was noted how lobbyists value winning, their image, and respect within the Washington community. This knowledge about agricultural organizations and their lobbyists, combined with an understanding of the complexity of the fragmented policymaking process within which lobbyists do business, demonstrates the importance of thinking about how relationships between lobbyists and policymakers are built and maintained. As can be seen in Table 3.1, representatives of agricultural interests employ a variety of direct and indirect means of building these relationships.

Lobbyists can hardly afford to be haphazard about whom they approach and how they go about establishing contacts with those who have policy responsibili-

TABLE 3.1
LOBBYIST STRATEGIES FOR GAINING
AND MAINTAINING ACCESS TO PUBLIC OFFICIALS*

USING DIRECT PERSONAL CONTACTS	USING INDIRECT CONTACTS OR INTERMEDIARIES
In an office visit, summarizing a problem	Arranging a meeting with members or firm officials
Attending a meeting on issue strategy	
Serving on an advisory board	Presenting research or technical information
Testifying at a formal hearing	
Informal social meetings	Employing a consultant with good political contacts
Doing favors in response to official requests	
Attending fundraisers	Activating grassroots lobbying
	Using another organization to present information
	Contributing PAC funds
	Planting a story with the media
	Litigating or threatening a law suit
	Employing a public relations firm
	Organizing a public demonstration

*Ranked according to lobbyist respondents' perceptions of usefulness

ties. In this regard, organizational representatives are limited by two externally imposed factors: (1) acceptance by the policymaker toward whom the group or firm's efforts are being directed and (2) the lobbyists' own perceptions about how others will judge the way lobbying is conducted. In other words, lobbying behavior depends as much on the actions and attitudes of those with whom the organizational representative must politically interact as it does on the policy needs of the group or firm being represented. This "transactional theory" of interest representation, first articulated by Bauer, de Sola Pool, and Dexter, suggests that group strategies and tactics—or the style of lobbying—may vary considerably under different political circumstances.[17]

There are few surprises about the transactions between agricultural lobbyists and policymakers. As was the case when Lester Milbrath first studied the Washington lobby, transactional relationships are based primarily on the exchange of information.[18] Agricultural lobbyists provide valuable information about farm, food, and foreign policy conditions, problems, trends, and issues that would otherwise not be immediately forthcoming in a useful form from either bureaucrats or academic specialists. That is, lobbyists provide interpretations of political conditions for policymakers who must avoid the risks of making policy decisions in uncertain situations. If anything has changed in the intervening thirty years since Milbrath's study, it has been mainly a technical evolution. In the mid-1980s, many organizations employ computer hardware, large data bases, and mathematical models to analyze complex issues quickly. The National Soybean

Processors Association, worried about U.S. trade policy and Argentinian competition, for example, can do a study of that nation's differential export tax system as it subsidizes Argentine producers. In order to address the need for countervailing duties before the Department of Commerce, the Rice Millers' Association can do a systematic review of Thailand's rice policies and their trade effects. Such analyses allow legislators and administrators to examine more closely variable conditions than was possible previously.

More striking a change has been the way in which lobbyists pursue access to policymakers. In both the Congress and the White House, staffs have grown enormously in both size and influence over thirty years.[19] Congressional employment grew by over 200 percent between 1960 and 1980, with the fastest growth rates in the 1970s. White House employment grew by about 50 percent during that time. In contrast, bureaucratic employment stabilized in size in the 1960s. These staffing trends have been accompanied by an increased reliance on employees by elected policymakers.[20] Although bureaucrats are no more able to buffer themselves by layers of subordinates than they were previously, elected officials can and do. As a result, policy information is almost always passed first to the staff member responsible for a specific issue area. Then the information is condensed and presented to a legislator or ranking White House policy adviser. Under such circumstances, it is highly unlikely that a sophisticated analysis will be studied in great detail by those with the greatest policymaking discretion.

Veteran Washington lobbyists, several of whom have been working in agricultural policy since the 1950s, describe this changing process of communication best. They remember small committee staffs; individual legislative offices with four to twelve staffers, whereas now there are from eighteen to well over a hundred; direct access to almost any legislator; and ease in contacting key presidential advisers. Contemporary conditions mandate that these same lobbyists—or their own staff—concentrate on getting information first to relatively lower-level employees. In Congress, it has become necessary to deal with both subcommittee staff and subject specialists from the offices of individual legislators whom may be interested for either constituent or committee assignment reasons. These exchanges are fairly routine as lobbyists struggle to maintain recognition by staff members, whom they see turning over in their legislative assignments with great frequency. One trade association representative follows four sequential rules for such encounters, which are typical of the agricultural lobby: (1) have something useful prepared to leave with the staff member, (2) be ready to explain its importance in no less than three minutes, (3) link the information as closely as possible to an electoral or constituent matter, (4) and try to arrange a later meeting with the legislator during which a one-page summary of the organization's position can be explained.[21]

What accounts for this change in approach? Long-term Washington lobbyists see four factors. First, congressional members are likely to sit on more committees and subcommittees than they were before, and legislators now pay more attention to the responsibilities of several subcommittee assignments. Second, there are far more lobbyists seeking access to each member of Congress. Third, more sophisticated data presentations, while useful, are also more difficult to comprehend and interpret. Fourth, with an increased congressional reliance on direct campaign efforts and a demise of the party as an electoral intermediary, legislators are less readily available to lobbyists. Members of Congress meet with constituents nearly every day. Then many legislators go home to work in the district or state on four-day weekends. Under such conditions, the congressional staff becomes indispensable in analyzing and understanding issues, reducing an overwhelming workload, and simply protecting the legislator. Things are not much different in the White House, where policy advisers are increasingly targeted for direct lobbying.

Staff-dependent lobbying does not mean that friendship-based lobbying of legislators and advisers is outmoded. It is simply less feasible and less routinely practiced, as lobbyists perceive it, than has long been the case. Younger policymakers who have entered the political process in a staff-brokered age are particularly unlikely to have many alliances with lobbyists. Personal relationships, which have resulted more likely from fund-raiser contacts rather than from leisurely dinners or excursions, generally have become more superficial. "I know far more people in [Washington] than ever," lamented one veteran lobbyist, "but I don't count many friends among them. You miss a lot."

There are several important implications of these changing conditions in terms of how the agricultural lobby is now organized. In each instance, organizations have changed their behavioral styles to reflect access problems that were far less evident in the 1950s and 1960s. For example, some groups have opted not to make routine contacts with members of congressional committees, but to use political action committee funds to gain recognition and entry when the situation warrants personal discussions. Some of these interests, in contrast, have chosen to confront and challenge established policymakers openly rather than to supply them with information cooperatively. The use of PACs is one tactical change used to threaten and coerce policymakers, especially by dairy and by California interests. Other organizations have broadened their operations and redefined their basic policy interests. Many turn outside the organization to hire consultants. These and other changes have brought different policy-relevant organizational features to the agricultural lobby that were not previously understood to be of consequence.

THE ORGANIZATION OF AGRICULTURAL INTERESTS

Agricultural interests choose their representational styles on the basis of one internal organizational factor and one external consideration. Members, officers, and staff must somehow decide what problems, issues, and public policies the organization will address. At the same time, and for as long as these political goals are on the organization's agenda, lobbyists must decide what sorts of relationships they will establish with various other policy participants.

In agricultural policy, twentieth-century farm groups historically have selected either broadly diffuse sets of political goals or narrowly intense ones. The general farm organizations, as observed earlier, followed the former approach in attending to any array of closely related farm-to-market considerations. As government moved to deal more with specific commodity programs in addressing farm problems, the single-crop and livestock groups emerged as effective representatives that followed that much narrower strategy.

Conditions changed, both in agriculture and in agricultural policy. Many of the commodity groups, as previously noted in Tables 2.1–2.6, represented both producers and middleman processors. As a consequence, their interests were not nearly so narrowly defined as was possible for all-producer membership groups. Then, too, agricultural problems became more encompassing for each commodity. Concern for prices and production practices was compounded by necessary attention to such matters as genetics, disease, marketing, foreign trade, and financing. Members of Congress were no longer able to differentiate between the general farm and commodity organizations on the basis of what issues attracted them and how effectively the groups were represented on either diffuse or intensive policies. By the late 1970s, some commodity groups were just as active as the generalists, and legislators saw them as even more influential on comprehensive farm issues.[22]

Beginning in 1977, complexity increased with the reappearance and new prominence of farm protest groups. The resulting political attention also brought many rural advocacy groups to the political forefront. The reform emphasis of these groups, with mainly a neopopulist ideology of redistributing political power to the grassroots populace, was even more diffuse than the policy concerns of the general farm organizations.

At the other extreme, the expansion of consulting enterprises brought increased specificity to the agricultural process. Once hired, consulting firms and their specialists are directed even more toward single proposals or issues than were commodity organizations in the 1960s. Consultants work not on a single interest but, instead, on a single project, even when employed by a large multipurpose lobby. Their time and attention is directed to making one thing work on a one-time basis. Other organizations with permanent professional staff have gotten involved in agricultural policy on a single-project basis as well.

Changes in the degree to which agriculturally relevant problems are represented politically as broad or narrow have important consequences for the political relationships and transactions that exist among those concerned with public policy. Established Washington lobbies have a continuing need to address the specific issues affecting their organizations and members (see Table

TABLE 3.2
A TYPOLOGY OF
CONTEMPORARY AGRICULTURAL INTERESTS

		ORGANIZATIONAL POLICY CONCERNS	
		Broad	Narrow
INTERACTION WITH POLICYMAKERS AND SUPPORTERS	Long-term	MULTIPURPOSE ORGANIZATIONS American Farm Bureau Federation National Council of Farmer Cooperatives National Cotton Council American Seed Trade Association Cargill Women Involved in Farm Economics	SINGLE-ISSUE ORGANIZATIONS American Farmland Trust National Peanut Growers Group National Soybean Processors Association Farm Credit Council ADM Milling National Food Processors
	Short-term	RURAL PROTEST GROUPS American Agricultural Movement Family Farm Movement North American Farmers' Alliance Prairiefire National Catholic Rural Life Conference Center for Rural Affairs	SINGLE-PROJECT ORGANIZATIONS Agricultural Research Institute Farmers for Fairness Ralston Purina Capital Legal Foundation Environmental Policy Institute Foundation on Economic Trends

3.2). As a result, these groups and firms will be returning to the same policymakers to look for both spokespersons and supporters of organizational positions. No matter how broad the policy spectrum, established interests have to provide some degree of ongoing attention to all of the policy components. Such continuity is far less a need for either the protest-style groups or for the special-project interests. These organizations can best direct themselves to more short-term relationships between policymakers and the organizations the lobbyists represent. As a consequence, the individuals pay far less attention to how the organization they represent will be treated when it enters the policy process. Therefore, a broader range of lobbying strategies and tactics may be employed as protesters, consultants, and others become willing to "pull out all the stops."

Differences in the extensiveness of lobbying efforts account for four types of private agricultural interests in national politics. Policy origins, redefinitions of political interest, transactional relationships with policymakers, and membership relations are critical in defining a policy type. From the broadest to the narrowest, these policy types include: (1) agrarian protest groups, (2) multipurpose organizations, (3) single-issue organizations, and (4) single-project organizations. Each type, along with its policy relevant differences, is described briefly below.

Agrarian Protest Groups

The organizations with the most encompassing interests in public policy are the agrarian protest groups. These groups desire sweeping reform in either food production or farm programs, and they adopt an adversarial style of participation. They are diverse in what they emphasize but alike in their general condemnation of widely accepted agricultural practices, and they eschew political bargaining, favoring instead strategies intended to force public officials into action. Many are farm groups, especially those organized during the late 1970s and throughout the 1980s. A large number, such as the Catholic Rural Life Conference, are more broadly interested in rural problems but share with the farm groups a desire to redistribute political power away from what they see as the Washington elite. The farm groups are organized around such policy ideals as price parity, reform of financial institutions, social justice, or family farm preservation. Some of their activists speak openly of the need for such long-range policy changes as Latin American–style land reform. Ideology, such as anticorporate beliefs, plays an important part in the policy discussions of these organizations.[23] Rural interests of this type are also ideological and organized around such high-impact reform issues as rural poverty and bans on chemical use.

The protest organizations tend to hold a strong and continuing commitment to the initial issues and ideologies around which the groups developed. Even in advocating a major farm legislation proposal, for instance, the American Agriculture Movement refuses to abandon applications of parity, even though it antagonizes policymakers.[24] Other farm groups, most notably the National Farmers Union, have moved away from this type of behavior and into another category by virtue either of achieving early political goals or modifying initial reform demands into more easily obtainable incremental policy changes. Continuing as a protest-style group, focused on a traditional goal such as preservation of the small family farm, provides an interest with a clear and distinct label. Such issue identification has the advantage of legitimizing the organization's claim to knowledge about a particular facet of agriculture, but it also places most of these groups at a disadvantage, since busy policymakers believe that they already know where each one stands. In addition, the pragmatism of most policymakers causes them to reject extreme positions; and they view protest groups as extremist. Therefore most protest groups have a difficult time gaining direct access to members of Congress and the administration.

Access to policymakers also presents a problem for protest-style groups because it is usually unclear who has jurisdiction over their broad issues. The rise of congressional subcommittees and federal agencies within the Department of Agriculture has divided agricultural legislation responsibilities into highly specialized components. Since policymakers in public institutions look for detailed information that is useful to their governing responsibilities, they are seldom open to lobbyists and activists who want to discuss major reform. The protesters and their demands are often relegated to special department projects with small budgets or to underworked subcommittees from which a small-scale or experimental piece of legislation may eventually emerge.

The transactional relationships between the protest groups and policymakers are generally brief, irregular, and limited. On the whole, these organizations suffer from a lack of familiarity and have no ability on a routine basis to discuss pending proposals with policymakers who structure them. As a result, most of the information and data that protest activists assemble gets channeled to the press or electronic media. The legislative- and administrative-policy spokespersons for reform issues tend to occupy policymaking positions, whereas they have not previously been accorded much status in agricultural matters.

The most consequential relationships that staff and officers of all interest groups have to maintain are the ones with those who provide the funding. Since many of the reform groups provide no services of any material use, retaining loyalty and affiliation is not easy. The strategy almost always revolves around providing the benefits of organization that attract and retain people who are or who can become interested in politics and then keeping them involved in group affairs. This leads to highly visible participants, relatively small memberships, and a decided reluctance on the part of group activists to discuss actual numbers of supporters. In terms of transactional consequences, it also has a strong bearing on these groups' ability to do much more than protest, since they lack so many traditional interest group resources. As a result, the greatest policy opportunity for the protest group lies in the hope that their activists will be able to link group goals to related but wider societal issues. Legislation supporting alternative systems of agricultural production, for example, was tied to massive publicity about chemical contamination. The high levels of price supports proposed in so many versions of the 1985 farm bill owed their presence to the ability of farm protest groups to turn national media attention over several years to farm strikes, tractorcades, forced farm sales, and foreclosures. Even in these successes, however, protest groups have done little to dictate actual legislative content.

MULTIPURPOSE ORGANIZATIONS

Multipurpose interest groups are defined by their involvement in numerous and widely different agricultural issues and by their heterogeneous members, supporters, and political contacts. They formulate policy statements and lobby on a wide array of legislative and administrative proposals. Although some issues receive priority consideration, lobbying takes place over a wide road map of national politics.

These interests include some of the general farm groups, several commodity groups, many trade associations, a few of the groups that represent both producers and marketing firms, and a small number of large agribusiness corporations. Examples include the Farm Bureau, National Cotton Council, American Seed Trade Association, and Cargill. In a given year a typical group from this category actively addresses general economic issues, agricultural financing, price supports, pesticide regulation, and worker safety legislation. The group also attempts to examine most farm policy proposals in order to judge the likely impact and chance for becoming law. Unless the organization is unusually well staffed, like the Farm Bureau or the Food Marketing Institute, the heavy work load normally precludes its lobbyists from doing much more than reacting to most issues. Support or opposition will be noted, a rationale given, and perhaps modifications suggested. In a few instances, complete alternative proposals may be drafted.

To compensate for the reactive situation that the multipurpose organizations encounter, lobbyists attempt to maintain regular contact with a wide range of policymakers in decentralized positions and, they hope, discover proposals at an early stage of discussion. This allows for greater opportunity to structure the direction of whatever legislative or regulatory proposals are being considered. As a result, multipurpose organizations place the greatest premium on—but cannot always hire—lobbyists who have ready access to individuals throughout Washington. For most of these groups their lobbyists collectively are seen as well placed to do the job if they "know everyone" or "understand the terrain."

The same behavior characterizes relationships between staff and members. Most of the organizations encounter strong membership expectations that all facets of farm or agribusiness life will be attended to in national politics. These are the organizations in which sweeping agendas emerge from membership meetings or corporate boardrooms. The staff must pay considerable attention to keeping the agenda setters informed about what issues and problems need to be addressed. The staff must also report on progress, explain new public policies, and frequently show why the organization lost on important issues. All this takes a great deal of time and attention and requires the use of several vehicles for direct communication with supporters. As a consequence, the multipurpose groups generally have the most impressive range of magazines, newsletters, flyers, conferences, workshops, and seminars of any of the agricultural interests. While the multipurpose groups constantly address new issues and problems, they are often better remembered for longer-term expressions of their beliefs about general farm or business conditions. Liberal, moderate, or conservative labels are applied to most of them by many policymakers, but these labels are modified by orientations to free trade, direct democracy, or some other world view. Given the pragmatic and reactive approach their lobbyists usually take to policy issues, such sentiments seem based primarily on the proclamations made through the organizational forums designed to involve their supporters. From them, members gain a consistent and stable image of their organization's political purpose; yet this is accomplished at some cost. Many policymakers are simply reluctant to get too close to lobbies that are seen as inflexible. The many organizations of this type, and the frequently contrasting viewpoints their images represent, have created a cautious policy environment.

SINGLE-ISSUE ORGANIZATIONS

Most observers point to single-issue organizations when they comment on the political advantages of specialized expertise, the need for routine access to only a few policymakers, and the group's ability to focus on one item of interest. This began to change in the early to middle 1970s. As noted earlier, many commodity and trade groups have felt compelled to become involved in a broad range of policy issues. This should hardly be surprising, since problems of imports, farm credit, medication for dairy cows, and so forth are nearly as important to milk producers as are their price support levels. The largest number of commodity and trade groups, several large agribusiness firms, and a few rural interests have chosen to maintain a single-issue lobbying style, however. Most of these organizations have made the decision to avoid other issues in order to maximize their limited lobbying resources and maintain their images. The single-issue groups and firms have been able to do so because neither staff nor membership expectations about appropriate group action have escalated beyond the initial issue around which the interest organized.[25] Smaller and more homogeneous organizations, for obvious reasons, have been best able to maintain this style of participation.

These are the organizations whose transactional political relationships can best replicate the subgovernmental model in which a few lobbyists usually work only with highly specialized legislative subcommittees and relatively small federal agencies. A prime example would be Sunkist, the grower-owned cooperative whose greatest political interest is in federal marketing orders for navel oranges. Lobbyists for such organizations attempt to see policymakers regularly, engage them in ongoing policy dialogues, establish relationships of trust, and gain the opportunity to discuss and even advance their own proposals early. This type of contact is especially important as lobbyists take time to address problems of changing industries. Groups that are the most successful at this strategy are concerned most frequently with problems of foreign trade or farm income maintenance. The same can be said of the clientele groups that have developed around USDA programs such as soil conservation, research, and reclamation.

Other groups that pursue such a strategy have been less fortunate. One of two reasons usually accounts for their lack of policy success. First, policymakers, for personal and political reasons, may not want to give access to a small organization that they may not know. Second, policymakers themselves may be responsible for small or out-of-favor programs that currently do little to address the problems around which the group organized. If noninfluential legislators and administrators are a group's primary contact with government, the interdependent status and limited opportunities of the interest to generate public support will prove to be a deterrent to any political goals. As was the case with the commodity groups prior to the greater fragmentation of power within subcommittees and agencies, many of the single-issue groups must simply bide their time and work for further structural changes in governmental decision making.

SINGLE-PROJECT ORGANIZATIONS

Single-project organizations have more sharply defined and immediate policy goals than those of the single-issue type. Single-issue interests have a continuous and comprehensive concern for legislation and regulations that affect a commodity or product. Single-product interests intervene even more intermittently, as their lobbyists see it, to promote a needed reform or to block disadvantageous rulings. Because of these circumstances, the organizations have the advantage of approaching policymakers with arguments about confronting an immediate situation. In many cases, these organizations' lobbying efforts may be made up only of consultants who work for one or more groups. Access to policymakers remains critical for either the organization or their consultants when they enter the agricultural policy arena over a current issue that both they and policymakers hope will soon be resolved. A prime example was the farm credit situation of 1985, in which both agricultural lenders and producers were overextended financially and needed quick relief. Only the Farm Credit Council was able to put together a lobbying team to pursue a policy solution for its members. Neither bankers nor farm interests were able to do so.

To be successful in reaching the ear of legislators and administrators, organizations of this type must be equipped with especially useful and persuasive information about the issue under discussion. Because the transactional relationships with policymakers are short- rather than long-term and dispersed throughout the agricultural arena, interest representatives cannot rely just on personal experiences and rapport with individual public officials that come from long-standing personal relationships. Instead most lobbyists from these organizations rely heavily on their own or their organization's reputed integrity or special skills. Consultants are sometimes hired to bring these often little-known organizations political access.

Not all single-project organizations are made up of or employ consultant lobbyists. Several foundations periodically attempt to influence agricultural policy through special reports, projects, and conferences. Similar strategies are used by staff members of several trade associations and agribusiness firms that only rarely become involved in any direct way with public policy decisions. Several interests more frequently identified with other policy arenas also intervene in agriculture in this way. In the recent past, these have included some religious, social-service, auto worker, and financial groups that employ agricultural staff specialists. Unless lobbyists work in coalition with those with a history and knowledge of agricultural policymaking, as they often attempt to do, their efforts are likely to be poorly received. Some of the trade associations, business firms, and other single-project group lobbyists, in fact, have reputations for being among the least effective representatives in agricultural policy. The complaint among policymakers is that, while these representatives know what they want, they "don't know agriculture or what's good for it." Interests of this type, it appears, have come to occupy a legitimate position in agricultural policymaking as long as their representatives can find some way to function with the finesse and knowledge of other, more established agricultural organizations.

These descriptions make clear how diverse agricultural interests have become in an attempt to deal with the expanded policy universe. There are several distinct sets of agricultural lobbies that not only represent varying policy interests but also operate in different ways. All are important. Because of what these interests choose to represent and how they do it, their impact on agricultural policy is no less variable. The next five chapters will analyze five major variations on the four types of interests discussed above.

4

Agrarian Protest
and a Grassroots Lobby

with Mark Lundgren

Cafe conversations are as much an enduring tradition of farm-town America as spring planting and fall harvest. Following the demise of those thousands of once socially active local farm groups, coffee klatches of farmers and businesspeople remain a forum for cussing and discussing agricultural policy and its effects on farm economics and rural communities. Since government programs are widely understood by farmers to be one of the three major determinants of farm income—the other two being domestic consumption and exports—this is not surprising.[1] They and their predecessor organizations may well be the cause or the effect of farmers ranking among the most politically active segments of the United States electorate.[2]

These small-town conversations of farmers and businesspeople usually influence little more than increases in personal body temperatures or changes in local opinions. Far more consequential, for both the participants and national agricultural politics, were talks held in Campo, Colorado, in midsummer 1977. For from these dialogues the American Agriculture Movement (AAM) emerged.[3] And from the AAM came a host of activist farmers who were socialized into organized group politics.

The participatory style that these farmers learned through the AAM was one of political threats, open confrontation with policymakers, and unwillingness to bargain over policy goals. With this style came political distrust and a widespread belief that both farm policymakers and established farm interest groups were unresponsive to farmers' needs. Although the AAM did not succeed in establishing lasting ties to many of these newly involved farmers, the movement did create a substantial leadership cadre who repeatedly formed local, statewide, and

64

regional opposition organizations throughout the 1980s as the protest style that historically characterized agricultural politics was reborn. These new organizations never became part of the political mainstream of Washington lobbying, but many of them remain involved quite heavily in the continued mobilization of farmers and the formulation of agricultural policy, filling what they see as an important representational void.

This chapter describes the evolution of mid-twentieth-century agrarian reform organizations as important components in the overall representation of farm interests. These organizations developed as a conscious attempt by resourceful farmers who were already represented politically to form an alternative interest group to shape agricultural policy. The reemergence of reform-oriented protest groups led to initially productive organizations, but their commitment to diffuse policy concerns made it impossible to develop well-defined policy proposals or to keep the organizations themselves from repeatedly falling into disarray. As a consequence, even though they generated positive responses from government, most other participants never saw them as playing a useful policy role, either in generating specific ideas or in facilitating a dialogue—as opposed to a monologue—between farm constituents and policymakers.

The new farm protest groups did not generate support in a political vacuum. Alienated farmers, as the earliest agricultural lobby, have often formed opposition groups based on populist and progressive ideals that sought more direct power for the farmer and better financial rewards for producers. In addition, there have always been at least a few other organizations of people who believed in some essential rural values and who continued to foster these ideals even when little or no political attention went their way. Several such groups were articulating their concerns about the direction of agricultural policy prior to the call for a farm strike.

Most of these were small groups, isolated but contentious in their public policy views. ACRES USA, for example, had been arguing for "eco-agriculture" as an alternative to chemical contamination for years. Its Missouri founder saw large-scale agriculture as a villain, and an exploitive national government as the villain's protector. The National Organization for Raw Materials (NORM), a scattered group of rural activists, argued that farm and national debts robbed raw-material producers of their basic goods. The Center for Rural Affairs in Walthill, Nebraska, was founded in 1970 as an advocate of small-farm life-styles. Center grants, in the early and middle 1970s, went to demonstration projects and the preparation of educational materials directed toward life-style protection. A variety of church-affiliated organizations, most notably Roman Catholic, addressed farm and food issues from the position of agrarian social policy reform. None of these interests were as yet involved in mobilizing a reform movement, however.

In its call for economic justice and its challenge to national policymakers the AAM gave common direction, a political strategy, and national attention to these generally ignored groups. After the 1978 tractorcade, the Center for Rural Affairs joined to bring other small farm-advocacy groups together in support of AAM-inspired legislation. NORM and ACRES USA representatives began to write and lecture on the need for parity, the AAM's primary political goal. Catholic church activists in 1979 turned from their previous interests in farm labor and world hunger conditions to address their support for family farms.[4] Even though the AAM did not create an idealized view of farming's place in the U.S. social and economic fabric, the new organization certainly turned that existing ideal back into a credible political demand by serving as a galvanizing force for a nation of citizens who believed in the superiority of a threatened agrarian life-style.

THE AMERICAN AGRICULTURE MOVEMENT

The discussions in Campo were reactions both to the immediate farm economy and to reports on the politics of the 1977 farm bill in Washington. The six originators of the AAM, who led the coffee shop discussions, saw agricultural policy decisions as remote, both physically and programmatically, from farm production. Colorado and adjacent high-risk agricultural production states were the ones especially plagued by the effects of falling commodity prices and escalating production costs. The region was especially troubled by increasing debt loads and failing farmers. Yet with the encouragement of President Carter, Congress passed a farm bill that reconfirmed incentives for high levels of production—and thus still falling prices—without what these farmers considered to be sufficiently high price supports.

Because their own financial difficulties were so evident, the Campo discussants could conclude only that the entire agricultural establishment of policymakers and Washington lobbyists was out of touch with reality.[5] The single strategy for ensuring policy responsiveness, as they saw politics, was to develop new farm representation through an organization that could express the anger of producers in some dramatic and attention-getting way. The farmers who met in Campo perceived a solution to an economic problem that they felt politicians had created. As they saw it, politicians rather than economic conditions made it impossible for producers to gain sufficiently high prices. With such a simple definition of what troubled agriculture, an equally simple policy demand was easily developed. AAM founders wanted prices federally supported at 100 percent parity levels and saw this as the only feasible policy response. To AAM leaders this meant federal payments for crops that would guarantee farm profits equal to average farm profits of 1910–1914, the traditional benchmark golden years of agriculture. To protect livestock producers, who received no price supports, the AAM also demanded strict limitations on meat and livestock imports. Expanding imports were also seen as a problem created by inattentive politicians rather than one defined by international economic conditions.

Although conclusions about the problem and its attendant policy solutions were drawn easily, the task of articulating this position was acknowledged by the Campo farmers and businessmen to be difficult. They did have faith, however, in the responsiveness of politicians and the public if their attention could be gained. The emerging cadre of strategists realized that one more organization that looked and sounded like those already in Washington could not get its views across, especially in competing with more established and recognizable interests. Moreover, coffee-time regulars argued, farmers were not easily organized around political goals, even if substantial organizational resources were available to use in attracting them.[6] So the chances of organizing an exceptionally strong and unique interest group seemed remote, until someone thought up the farm strike of 1978.

From the junction at which policy goals and an organizational plan were brought together, events moved very quickly. It was agreed that the new group would prompt and assist farmers throughout the country to organize as local groups, much along the lines of Farm Bureau county chapters. The locals would be pockets of farmer interaction and discussion that would inspire political activism directed toward a national strike of all food and fiber producers. The local organizations would join statewide and national organizations in massive demonstrations to protest farm conditions and advertise the strike. This strike would go on until federal law backtracked from the 1973 and 1977 farm bills and accepted price parity.

As the first step, the six organizers called a meeting of local farmers in early September 1977. The initial supporters met to form the core of an ever broadening series of similar gatherings that, in turn, would become the basis for the hoped-for local organizations. Approximately forty residents attended that first meeting and were presented with the outline of a strike as well as the newly formed group, the American Agricultural Strike (AAS).

The local participants were asked at the first meeting to provide the manpower to arrange a larger rally in the county seat of Springfield to draw up specific strike plans and present them after only one week. The short time frame was intended to reinforce the urgency of economic conditions and, through a flurry of participant support, to create the necessary organizational momentum to spread the movement. The use of flyers, personal calls to potentially sympathetic friends, and appeals for media coverage brought out an estimated seven hundred farmers in Springfield. Most areas of Colorado, Kansas, and Texas were represented at the meeting. Participants were asked to observe how quickly a large and supportive turnout could be gathered. This support, rally speakers argued, demonstrated that farmers were angry enough, smart enough, and committed enough to band together politically on a national basis if organizers did their work. After accepting a December 14 strike date and establishing a national farm strike headquarters in Springfield, participants went out to organize the rest of the country. Two weeks later, approximately two thousand protesters attended a national rally in Pueblo, Colorado. Much to the strike leaders' delight and good fortune, national network television reporters and U.S. Secretary of Agriculture Bob Bergland also attended. As a consequence, politicians and the public soon gained awareness of agrarian frenzy and a threatened production boycott, just as the Campo discussants had hoped.

The strike movement of 1978 took two distinct organizational directions, one decentralized and the other centralized, which were to be politically consequential for the next several years and into the 1980s. The nationwide system of locals was modeled on the successful Campo/Springfield rallies. Local pockets of activism were encouraged wherever strike enthusiasm and peer pressure could promote turnout. Facilitators from headquarters and other activist sites traveled extensively to keep the contagion effect of escalating protest in public view. By strike deadline, forty states were organized and represented at a national American Agriculture Movement convention. Over eleven hundred locals were reported open by January 1978, and more than three million people had attended AAS/AAM rallies by then.[7]

These organizers and supporters were learning grassroots politics as a consequence of their local work. While most AAM participants had membership experience with other farm groups, they were now learning about something different from staff-assisted, democratic agenda building within an organization such as the Farm Bureau.[8] By emulating civil rights and antiwar activists, their own agrarian forefathers, farmer protesters learned that seemingly unresponsive politicians and bureaucrats would turn out in their own farm communities in the face of anger, hostility, confrontation, disruption, and threats.[9] These activists also came to believe that policymaker attention indicated that farmers were viewed as politically important. They saw themselves gaining political legitimacy on the basis of their own individual claims as these claims were expressed in an organizational setting that farm protesters saw as diametrically opposed to the behavior of Washington farm lobbyists. Events during the first half of the 1980s, which led to continued protest-style mobilization, indicated that these beliefs and feelings of farmer efficacy transcended affiliation with the AAM or identification with the farm strike.

The centralization of the AAM also began immediately after the Springfield rally and proved no less important than decentralized farmer mobilization. Organizers originally believed that fragmentation in locals could preclude a common purpose. Early organizing experiences in developing state units reinforced this belief in the need for a strong headquarters role in creating both consensus over articulated policy demands and sets of common activities that could be identified as characteristic of the AAM.

This leadership role was unlike that of contemporary farm interest groups with their organizational hierarchies, specialized staff assignments, lobbyist-directed grassroots activism, chains of command, and governing rules. The AAM initially eschewed membership dues, rolls, officers, and rules, but AAM leaders still developed a highly centralized organization with many specific responsibilities delegated to early activists. Headquarters efficiently processed paperwork, made and monitored assignments, kept track of organizers and locals, maintained a master list of many of those who attended AAM meetings, and ensured daily routine operations. In addition, a functional—if not formal— division of labor emerged within the first days of operation. Volunteer staffing, recruitment, fund raising, political liaison, economic analysis, public relations, and publications were carefully coordinated by a management team composed mostly of the original organizers. These individuals, as group managers, knew that protests and strike threats would be insufficient in explaining policy goals to those who would have to act on them. Supplementary information about farm conditions, the AAM's proposed bill to return to 1950s-type commodity program supports, and the political ramifications of ignoring the movement were also considered essential.

Because of these beliefs about a centralized lobbying strategy, the AAM organized effectively to get explanatory materials directly to both policymakers and the media. In order to overcome regional and commodity diversity among farm supporters that could lead to messages confusing to policymakers, organizers in Springfield also devoted considerable attention to what members might say. This entailed integrating a communications network among the states, focusing protests on a national tractorcade as a culminating experience for the AAM, determining what messages participating farmers needed to send from their locals, and explaining the best means for transmitting and reinforcing these messages.[10] The result was an orchestrated mobilization of constituent supporters that rivaled in method, though certainly not in behavior, the best techniques of any Washington lobby.

The lasting impact of the AAM's centralization has been twofold. Although its second strike failed and its second Washington tractorcade offended the nation, the organized American Agriculture Movement carried on as a Washington interest group long after most 1977–1979 participants had discontinued their support.[11] As a Washington lobby with strong ties to grassroots activists, the AAM maintains a unique and dissenting farm voice favoring high payments to farmers. Almost alone, the AAM popularized and then kept the issue of a "farm crisis" alive in Washington from 1978 into the 1980s.

The AAM's second contribution, in terms of its centralizing of organizational functions, has resulted from the leadership skills and appreciation for organizational entrepreneurship that former AAM activists bring to newer farm protest groups. These organizations, mostly local and state, have proliferated widely since 1983. Most have shown strong leadership both in their political strategies and in organizing their groups around farm problems. As a result, these organizations are politically effective, know how to influence supporters, and work hard to keep leaders and the rank and file together. In addition, the leaders of these organizations want to avoid the problems they eventually saw in the AAM.

A SPLIT IN THE RANKS

Perhaps the American Agriculture Movement's biggest problem was with a farm strike that even its leadership knew could never work.[12] As organizers in 1977, however, the AAM's leadership had no options other than the strike if they wanted to develop an alternative interest group. They could only hope that policymakers would acquiesce before planting began. Nonetheless, the failure of the strike and its political goals resulted in the first wave of massive disaffection within the AAM.

The plan for creating an alternative farm representative in national politics demanded a vehicle for rapid mobilization of supporters. The strike, with its calls for momentum and sustained commitment, was an excellent organizational cornerstone. Participants in the protests and demonstrations were not committed to a costly boycott, however. In fact, as the organizers understood, there were to be four distinct levels of involvement in the new group. At the first level, the AAM could count and advertise as its supporters the total number of participants at meetings. The second level of involvement meant informal identification as an AAM member. Members could also decide to assume specific duties within the organization. Actual boycotting was again a separate and nonmandatory decision. Though these options provided the leadership with certain advantages, unfortunately for the AAM, supporters were not in a position to understand these dynamics. As might be expected, there were constant complaints about peers within the emerging group failing to do their part. As responsibilities grew in each state and local unit, the burden of organization fell on increasingly fewer people.

Legislative successes in early 1978 did not soothe internal frustrations or the anger that the AAM had aroused toward policymakers. Since these successes were substantial, bitterness must have been deeply rooted. The January tractorcade had initially been well received as official Washington opened its doors to the unusual spectacle of protesting farmers. Because of the AAM's obtrusive presence, Congress passed the Emergency Agriculture Act of 1978 and the Agricultural Credit Act of 1978. Several supportive executive actions were taken within the USDA as well.[13] However, neither the AAM's proposal to bring about 100 percent parity through higher supports for commodity programs nor its cost of production plan was accepted. AAM leaders, openly and in private, interpreted this as total rejection. On the one hand, AAM leaders saw these events as the work of U.S. agriculture's "enemies" in the international financial community operating through President Carter. On the other hand, these activists argued that farmers had not done enough, did not remain united, and would have to come back to Washington better prepared. These divisive feelings immediately began to split the AAM apart, eventually robbing the farm protest movement of a clearly identifiable national voice.

Members first began to drift away from the AAM in 1978 after the legislative initiative had failed to bring about a victory for parity. The split over support of the cost-of-production proposal led many to desert the AAM at this time. The second tractorcade in early 1979 hastened the exodus. Several of the 1979 tractorcade organizers felt the need for a more forceful expression of farmers' views. In assuming command, these individuals moved to create traffic jams, illegally occupy the Mall in central Washington, and generally wreak havoc. The extensive damages turned many previous supporters against both the AAM and agricultural protest as a political strategy.

In the early 1980s the AAM was floundering. One faction emphasized a Washington-based staff lobby that could work inside Washington and also organize grassroots electoral involvement. In 1980, leaders of the AAM campaigned for Ronald Reagan in thirty-four states in an attempt to develop favorable relations with the White House. A second faction continued to emphasize grassroots activism. One subfaction of these locally oriented activists was convinced that policymakers would never respond to the gimmickry of national protests. It sought to reach legislators back home. A second subgroup within the grassroots faction felt that farmers should abandon direct politics in favor of educating those in local communities about their economic dependence on agriculture. Over the next few years all of the grassroots supporters remained irritated at headquarters activists for supporting Reagan, a president who quickly voiced a free-market approach to agriculture that was in direct contradiction to demands for parity. This bickering, and continued embarrassment from isolated outbreaks of violence associated with the AAM, further reduced the ranks of supporters.

The organizational split between those directed toward national politics and those favoring a local approach came to a head over a proposal by a Washington consulting firm to direct the public relations efforts of the AAM. Martin Haley Companies, a firm with considerable expertise in agriculture, came forth with a plan to enhance the political image of the AAM. Haley wanted the AAM to form a Political Action Committee, use the firm to "coach" its lobbying, abandon protest for all but symbolic purposes, and institutionalize a formal dues-paying organizational structure.[14] By 1982, as conflict over this proposal intensified, the AAMs had become two separate organizations. The Haley viewpoint and the national office were represented by AAM, Inc. Grass Roots AAM, headed by one of the Campo founders and championed by the official newspaper *American Agriculture News,* remained loyal to informal organizational arrangements and a protest style of political opposition.

From that point on, there were two clearly identified organizations, several points of view about political strategies, a single policy goal of price parity, and so much organizational noise that further legislative proposals were left unidentified within either faction. As a source of even general policy ideas, the AAM remained vacuous for many years.

It was not until planning stages for the 1985 farm bill that the AAM could put forth a policy agenda. Even then it was a loose collection of ideas advanced by many splinter groups associated with the AAM. Ideas included mandatory production controls decided by a producer referendum, parity-based loan rates, export subsidies, import labeling, the rescheduling of farm debts, the elimination of tax loopholes, reductions in commodity futures trading, and the establishment of a producer's policy board within the USDA.[15] The AAM was unable to consolidate these items and draft them in the form of an alternative farm bill proposal, however. With over a dozen such proposals before Congress, the AAM was reduced to the role of a policy dissident attacking nearly every segment of American agriculture. Other protest organizations promoted their own plans. For those who had envisioned new farm representation and then pursued the political goals of AAM, Inc., 1985 was another bitter experience.

The AAM was not disenfranchised as a proponent of policy alternatives only because its own ranks were split. The grassroots activism that had won so much favor and support within the AAM led many activists to start their own independent groups. New organizations were especially attractive to those interested in local activism, farmer services associated with the farm crisis situation, and organizational independence from AAM rivalries. By 1984, the AAM's inability to satisfy its constituency of reform-oriented producers was resulting in a bewildering proliferation of new interest groups. Because they produced a strong but only general impression of extreme farm discontent, the policy impact of these organizations forced policymakers to consider expanded income maintenance programs and a retraction in some export efforts. Like the AAM, however, the new groups did little to promote specific alternatives for policy change. They simply moved the focus of farm protest further from a centralized leadership and into the newly active, more recently hard-pressed states of the upper Midwest.

PROLIFERATING ORGANIZATIONS
AND FARMER SERVICES

The new organizations also capitalized on two incentives that the AAM had missed in maintaining its supporters: personal service and farmer assistance. Despite impressive shows of solidarity, many former members of the AAM who went on to organize other protest groups expressed an uneasy feeling that the movement "never cared about any of us as individual farmers." Conflict over centralization, the involvement of Martin Haley, the PAC controversy, and the Reagan campaign effort only intensified these feelings and led to a charge that the Campo founders could not have predicted in 1977. "Farmers needed help," said one veteran of the national tractorcades, "and these guys decided to go out and give their help to politicians."

It was in this context that a new set of farm protest organizations formed. Unlike the AAM, the new organizations did not direct their efforts immediately toward a nationwide program for achieving higher commodity prices. Instead, most of them joined around a program of providing services to help local farmers cope with the financial and emotional distress that resulted from the farm debt crisis. In time the new organizations developed an impressive array of member, or constituent, services that eventually provided foundations for more stable protest organizations than AAM locals. The membership incentives provided by these constituent services were crucial because, like the AAM in 1977, the new organizations lacked the financial resources necessary to offer selective material incentives commensurate with those routinely provided by long-established farm organizations. The service program also conferred legitimacy on the new organizations and drew favorable publicity from the media and praise from community leaders. This allowed some of the new organizations to become effective contenders for political influence without first having to engage in a massive show of strength. The service programs also enhanced opportunities for the new organizations to win contributions from churches, universities, foundations, community groups, and local governments. Such contributions came in the form of money, expert assistance, and volunteer labor, giving many of the new organizations the resource base necessary to maintain a high level of organizational activity from 1983 to 1987. The organizational benefits derived from the service programs also allowed the construction of locally oriented, farmer-run organizations that proved capable of mobilizing widespread grassroots support for the national political agenda carved out by a coalition of farm protest groups.

The programs created in Iowa are typical of those developed by the groups that followed the AAM, except that the Iowa Farm Unity Coalition assembled one of the most comprehensive sets of services. The coalition was formed in 1982 by state farm leaders to forge a coordinated response to the impending farm crises. Initially, it did not offer formal memberships to individual farmers. Individuals participated in the coalition by joining one of the affiliated organizations. Eventually many farmers came to participate in coalition activities without joining an affiliated organization such as the state chapters of the AAM, NFO, or NFU. In response, coalition memberships were offered directly to individual farmers. However, many farmers continue to participate in coalition activities without becoming formal members. For all practical purposes, then, the Iowa Farm Unity Coalition is a loosely knit, locally controlled network of farm activists. Its board of directors provides a common agenda for action but has been unable—and perhaps unwilling—to impose rigid discipline on the rank and file. Consequently, the coalition's efforts to mobilize farmers is only sporadically successful. Participation rates at coalition events vary greatly according to the nature of the activity, the timing of the event within the crop cycle, and the general level of morale and efficacy.

Many local organizations have helped the coalition. A deliberate effort was made to involve local ministers. Iowa churches, which have a tradition of social activism, have contributed money, labor, and facilities to support constituent services. Coalition organizers frequently use church networks to extend their reach.

Iowa's labor unions have been another source of support. Because much of the industry in Iowa manufactures products for agriculture, local unions joined in the coalition's efforts to improve the farm economy. The United Auto Workers, whose Iowa members work primarily for agricultural implement manufacturers, have been particularly supportive of the coalition. The coalition has also worked with the local citizen action groups to secure new benefits from state government for farmers. The work of Prairiefire Rural Action has been of special importance to the coalition. Prairiefire originated as the Midwest office of Rural America, but after years of operational autonomy it incorporated as an independent organization in February 1985. Prairiefire provides office space and staff support for many of the coalition's member services. In fact, the two organizations work together so closely that they are often confused or referred to interchangeably. The various supporting organizations have provided the coalition with much-needed stabilizing resources. They have also been valuable allies when the coalition has contended for political influence within Iowa.

The centerpiece of the Iowa Farm Unity Coalition's member services is its advocates program; and the central component of the advocates program is the coalition's Farm Survival Hotline. The hotline does not dispense disinterested advice. Operators are there to take the side of the farmer and to do everything possible to provide aid. The coalition has worked with organizations in several states—including other grassroots farm organizations, rural advocacy organizations, universities, and agencies of state government—to develop its expertise in advocating for farmers. The hotline, which has become a national model, has achieved a solid reputation for expertise and successful advocacy among both friends and opponents.

The volume of calls that comes over the hotline has been too large to allow the operators to give adequate attention to the problems of the individual caller, so the coalition built an advocate network to aid in processing the caseload that the hotline brings. The advocates are generally farmers whom the coalition has helped with a financial program and then educated about farm debts and foreclosures. Their function is to help farmers begin to assemble a strategy for dealing with their financial troubles. To the greatest degree possible, the coalition attempts to match farmers with advocates who have had similar experiences. If the farmer faces a Federal Land Bank foreclosure, the coalition tries to send an advocate who was also foreclosed by the Federal Land Bank. Having face-to-face help from a farmer who has gone through the same ordeal is often an important source of social and emotional support to a farm family in trouble. The advocates also perform several tasks to aid farmers in obtaining legal help. These include assembling all the relevant loan documents, helping with cash flow projections and other financial statements, finding an accountant who understands the tax consequences of various alternative courses of action, and acquiring the service of a competent lawyer. The staff of Prairiefire provides additional services to advocates to aid them in their work. Among them is the legal referral service.

Prairiefire periodically surveys Iowa farmers about the performance of their lawyers and has drawn up a referral list based on the findings of the surveys and the reputation of the lawyer. Although the advocates' network is geographically uneven and advocate burnout is high, face-to-face help has been provided to hundreds of farmers.

The Iowa Farm Unity Coalition has several other components in its service programs that link coalition leaders to the grassroots. It holds frequent meetings around the state to disseminate information about debtors' legal rights. The meetings usually consist of a lecture on the basic laws governing loan contracts and foreclosure proceedings, followed by a question-and-answer session. There is also a table that displays pamphlets concerning farmers' legal and financial problems. The literature is generally produced by the staff of Prairiefire, although the coalition also exchanges literature with other service-oriented farm organizations. At such meetings the coalition is not hesitant to present its political agenda, but politics is secondary to the business of educating farmers about their legal rights. Frequently the coalition speaker will ask to schedule another meeting with the same group to discuss farm policy and present the coalition's political agenda more completely.

The coalition has not been content simply to hold periodic meetings around the state. It has attempted to establish an organized presence in more than half of the ninety-nine counties of Iowa by forming county Farm Survival Committees. These committees meet regularly to plan local activities. Much discretion has been left to local leaders to determine the course a county committee will take. While some aggressively pursue the protest agenda of the coalition, others concentrate on advocacy work or serve as emotional support groups. Most of the groups have demonstrated to the local public and media the extent of the crisis in their own counties. One of the most popular tactics for doing this has been to plant a white cross on the county courthouse lawn for each local farm or rural business lost during the farm crisis. As with the advocates' network, the burnout rate among county group leaders is high. Consequently, this program has been only somewhat successful in building ongoing committees.

The coalition has two social-service programs aimed at aiding farm families in distress. First, it has worked with local ministers and social workers to help organize dozens of farmer-to-farmer support groups in which financially distressed farmers can air their emotions and get counseling. Second, the coalition has worked with the Iowa Inter-Church Agency for Peace and Justice to get food stamps for financially distressed farm families and to help farmers overcome their inhibitions about using government social-service programs. Part of this campaign was a lobbying effort directed at modifying eligibility requirements that penalized farmers by emphasizing their assets rather than their net incomes. The other part consisted of group sign-ups to lessen the stigma farmers may feel by enrolling individually.

Finally, the coalition has cooperated with the staff of Prairiefire Rural Action and others to help organize two programs directed at increasing farmers' knowledge of and influence over key institutions. The Bank Closing Response Team was developed to inform farmers of the problems they are likely to face if their bank closes. Between April 1982 and August 1986 twenty-seven banks closed in Iowa, primarily because of the depressed farm economy. Depositors at closed banks are protected by the Federal Deposit Insurance Corporation (FDIC), but farmer borrowers often find that bank closings jeopardize their financial situation even if they have kept their loan payments current. This is because many of the solvent banks that purchase the failed banks are reluctant to assume agricultural loans, for fear that such loans will eventually fail, thus threatening the newly reorganized branch bank. Since new sources of credit are not easily acquired under depressed economic conditions, many farmers lose their access to credit when their bank closes. The bank failure response team informs farmers about FDIC procedures and prepares them to face the credit problems resulting from their bank's failure.

The Farm Credit System (FCS) Task Force was organized to help FCS borrowers use their membership rights to change Federal Land Bank and Production Credit Association (PCA) policies regarding forbearance and foreclosure. The task force has worked with Iowa's U.S. senators to make the Farm Credit System more accountable to farmers. Although the task force achieved notable successes in changing the policies of a few local PCAs and land banks, it has been unable to mount the massive statewide effort necessary for a dramatic effect on the FCS practices. Because the fate of the Farm Credit System has become a national issue, the FCS Task Force plans to direct its future efforts toward influencing federal policies.

The service programs confer substantial benefits and impose no less substantial resource costs on Iowa Farm Unity. Salaries must be paid to hotline operators and support staff, materials for advocates and county committees must be updated constantly, and new activists must be trained. Although the coalition has been fortunate in receiving generous contributions of money and labor from churches, unions, and through Prairiefire, mobilizing the resources to maintain the service programs is a continuous struggle. Effort must also be made to extend the geographic reach of the service networks, and because of the higher rate of burnout among activists, the networks continually need to be rebuilt. The caseload brought in by the advocates program periodically overwhelms activists and drains effort from the coalition's political programs. For these reasons, the more politically oriented coalition members have sometimes questioned whether the service programs justify the costs. There is special reason to raise this question about the hotline because it is not used extensively to recruit for the coalition's political campaigns. Callers' names are recorded for future reference, but flyers are not sent out routinely. Callers are often told about coalition activities in their areas, but they are not placed on the mailing list unless they request it. However, the hotline and the other service programs do produce several important benefits for the coalition and hence make its political work more effective.

First, by providing a service of great value to financially distressed farmers, the coalition has achieved legitimacy in the eyes of many farmers, community leaders, and politicians. The advocates program has been effective in provoking established agricultural and political interests to develop similar programs. The legitimacy and goodwill derived from the advocates program are of great value as farm activists contend for political influence. The coalition has exercised disproportionate influence precisely because the advocates program created the perception that the organization is the legitimate representative of the farm community most affected by the farm crisis.

Second, the service programs have brought the coalition much favorable attention from the media. The degree of media attention drawn to the advocates program is well illustrated by the fact that the chief hotline operator found it necessary to place a sign on his door to indicate whether he is "on the air" or "off the air." Farmers involved in the group food stamp sign-ups were encouraged to call a press conference to dramatize their desperate financial straits. The irony of Iowa farmers signing up for food stamps provided a story angle the media could not resist. The coalition is also skillful at exploiting the symbols and images of what one staff member calls "agrarian fundamentalism." Activists garnered favorable publicity when they carried their white-cross protest into the governor's state-of-the-state address. The publicity derived from the service programs combined with skillful construction of media events allowed Farm Unity to acquire a high profile without having to amass a substantial resource base or a large professional staff.

Third, the hotline has been valuable to coalition leaders as a source of information about conditions and sentiments in the countryside. Information acquired over the hotline lets coalition leaders anticipate emerging problems and prepare a response. It also helps them cultivate contacts with politicians and reporters eager to keep informed of emerging trends.

Fourth, the advocates program provides durable, though intangible, membership incentives of the forms Peter B. Clark and James Q. Wilson describe as purposive and solidary.[16] Purposive incentives are derived "from the sense of satisfaction of having contributed to the attainment of a worthwhile cause" and are always dependent upon "the stated objectives of the organization."[17] Unlike those of the AAM, the objectives of the Iowa Farm Unity Coalition are not only to win new benefits in Washington, D.C., but also to provide practical help to farmers in financial distress. Coalition activists point to many instances in which they have provided aid to fellow farmers. The continual, concrete sense of accomplishment derived from such work is an important incentive to continued participation in the coalition. Solidary incentives, which may be individual or collective, are intangible rewards derived from affiliating with a group. Farmers in financial distress often feel stigmatized and may find that friends and neighbors shun them. Observations in Iowa create the strong impression that the social acceptance and emotional support received in county Farm Survival Committees and farmer-to-farmer support groups are effective incentives. The opportunities for exercising grassroots leadership made available by such groups is also an important attraction for many long-time farm activists.

Efforts similar to those of the Iowa Farm Unity Coalition that use constituent services to build a sustaining resource base have been undertaken by farm organizations in most midwestern states and in several southern states as well. Because of their common political purpose and problems, these organizations have coalesced into a new social and political force in rural America as regular interorganizational meetings are held to share information and experiences concerning the operation of the service programs. In 1986, participating state and local groups established a formal political alliance under the auspices of the National Save the Family Farm Coalition (see Table 4.1).

Whether these second-stage farm protest groups can succeed politically as they have done organizationally remains problematic. They have some obvious assets. First, many of their members are highly involved in the organizations and therefore are willing to work continually toward achieving the goals set by the leaders. Having benefited from their free and helpful services, many nonactivist farmers feel a significant sense of obligation to these new farm groups. Resources of volunteer assistance and money are often contributed by such supporters and are available from nonfarm supporters as well. Purposive and solidary incentives derived from the service program keep at least a core group of activists working hard locally even during periods when events in Washington are demoralizing. The political efforts of these second-stage agrarian protest groups, whether directed toward a policy or electoral end, provide a strong organizational base in over one-third of the American states.

Second, the new grassroots groups have developed genuine expertise in addressing many of the problems financially distressed farmers are likely to face. This expertise gives them legitimacy as representatives of those farmers most threatened by the financial conditions of the farm crisis. While maintaining their protest identity, many of these groups have used their expertise to achieve substantial political influence without having either to mobilize enormous stockpiles of resources or to organize large AAM-style rallies and tractorcades. Furthermore, a grassroots organizational structure exists to mobilize new service projects as the need arises, as it did to provide drought relief for the Carolinas.

Third, unlike the original AAM, these organizations have learned to work with many government agencies and state policymakers. Along with their community activism, this involvement has made the farm crisis groups respected and credible, at least locally and at the state level.

There are liabilities, too, however. First and foremost, at the level of federal farm policy, these protest organizations still occupy, in the words of resource mobilization theory, challenger status.[18] They have neither routine access to

TABLE 4.1
NATIONAL SAVE THE FAMILY FARM COALITION (1986)

American Agriculture Movement, Inc.	Minnesota Groundswell
California Agrarian Action Project	Minnesota Northwest Coalition for
Community Farm Alliance (Kentucky)	Rural America
Dakota Resource Council	Missouri Rural Crisis Center
Family Farm Organizing Resource	New York State Farm Alliance
Center (Minnesota)	North American Farm Alliance
Farm Alliance of Rural Missouri	Educational Project
Farm Crisis Committee (Nebraska)	Northern Plains Resource Council
Federation of Southern Cooperatives	Powder River Basin Resource Council
Illinois Farm Alliance	(Wyoming)
Illinois South Project, Inc.	Prairiefire Rural Action (Iowa)
Iowa Farm Unity Coalition	Rural Virginia, Inc.
League of Rural Voters Education	Texas Corn Growers
Project (Minnesota)	Western Colorado Congress
Magic Valley Borrowers Association (Idaho)	Western Organization of Resource
Michigan Farm Unity Coalition	Councils (Montana)
Minnesota Citizens Organizations	Wisconsin Farm Unity Alliance
Acting Together	

Washington decision makers nor established claims to exert influence over particular government policies. Although they have a committed membership, small state and local groups do not have the resources or the enthusiasm to employ professional staffs and lobbyists commensurate with those of the more established agricultural interest groups. Their ambitious agenda of credit reform, debt restructuring, production cuts, and loans for planting makes demands Washington policymakers are not used to hearing, much less accommodating, from recognized farm interest groups.

Second, the history of political action by the groups to which supporters actually remain loyal has been characterized by only sporadic cooperation. Sending hay to the Southeast is a different form of collective action from forging a national coalition capable of winning new benefits for farmers in Washington. The immediate genesis of many of the state groups came in 1983 when Jim Hightower and his associates began to meet with those in the existing network of farm organizers to put together a set of proposals for the Farm Reform Act. Hightower succeeded in assembling the basis for a farm bill proposal, but it was the result of diverse and unrelated policy demands from scattered farm activists that protest leaders put together as a bill in 1985. As a result, the groups had a great deal of difficulty in developing an agenda for comprehensive agricultural policy reform. It was not until 1987 that the groups addressed the many other components of farm and trade problems. Even then, cooperative efforts were limited because many groups continued to alter their ideology and goals in seeking·a specific policy identity.

Local, state, ideological, and crop differences limit interorganizational political cooperation and protest coordination. Group leaders have not always set aside personal differences for the sake of unity in the coalition, and organizational turf fights threaten the very existence of the national coalition. The groups have received substantial contributions of resources from nonfarm sources—Willie Nelson's Farm Aid being the most significant—but it is unclear whether these outside resources promote unity or exacerbate intergroup competition for funds. Many geographically close groups have even refused to combine services for the benefit of constituents. The groups, however, have been able to work together to draft, refine, and resubmit a farm bill proposal, but not without significant conflicts. Moreover, they have not succeeded in acting as a unified force in lobbying for their bill in Washington, and without such unity they are unlikely to achieve political success at the national level.

A STRATEGY AND IDEOLOGY OF PROTEST

When agrarian protest was reborn in 1977 as a political phenomenon, AAM activists were angry but nonetheless pragmatic in their demands. They had studied the political scene and felt that they had a management plan for achieving an attainable goal. To an economic analyst, parity may seem unrealistic and impractical. Critics of parity pricing contend that artificial maintenance of commodity prices to reflect the favorable cost/price ratios of previous decades would distort market factors, raise food prices, reduce American competitiveness in unstable export markets, and, by ignoring the greater growth in the productivity of agricultural resources, guarantee farmers a larger return on capital and labor than is enjoyed by other economic sectors. Farmers, however, do not have to see economic conditions that way in order to press resolutely for what they want. Serious congressional consideration of a flexible parity plan in 1978 demonstrated that several policymakers, for whatever reasons, agreed with the legitimacy of the AAM's parity demands.

Pragmatism, in the face of escalating anger and feelings of political rejection, was not easily maintained. Broad reform goals without specific policy proposals heightened this problem. The ideological arguments that the AAM used in mobilizing its strike supporters further helped undermine the always reluctant willingness to work with policymakers for change. With the help of representatives from organizations that were also opposed to existing agricultural policy, the AAM had developed an anti-establishment rhetoric that explained current farm plight and elevated parity to the status of key determinant of national economic well-being. The National Organization for Raw Materials and ACRES USA were relied on most heavily for their macro-economic theories of oppressed farmers. At first, the AAM's rhetoric merely glorified the family farm, agrarian values, and the contribution of agriculture as the underlying basis for general economic security. Agriculture's enemies were seen as uninformed and inattentive.

After the AAM's parity demands were rejected, this rhetoric hardened. High production costs were blamed on corporate conspirators. "Corporates," activists argued, ruined the farm economy through monopolistic handling of agricultural products. Bankers and other international interests, the "master swindlers," were behind the corporations and controlled agricultural and economic research, thus avoiding public exposure. Welfare economics, propaganda, and tax load victimization of working Americans were portrayed as tools of the internationalists, who sought to undermine America's moral fiber. The Trilateral Commission, David Rockefeller, and Jewish financiers were singled out as especially important conspirators.

In developing a protest ideology, the AAM and its spin-off organizations only followed the lead of every other American agrarian movement that had achieved national political prominence.[19] The AAM also created a well-defined niche for these organizations on the fringes of agricultural politics, a niche identified by policymakers and established interests as exceptionally radical in policy.[20] As organizational tensions consumed the AAM's ability to work with others on specific policy proposals, and as AAM leaders remained hard and fast in their opposition to negotiated agreements, the radical label became unshakable. Violent protest and examples of civil disobedience, such as Wayne Cryts's fabled raid to free his soybeans from a bankrupt grain elevator, only added to the image.

By 1980 modern farm protest had gained a uniform reputation of ideological posturing, even though many farm activists later sought to act more moderately. Policymakers and representatives of other interests, who could not see specific points of agreement with the protesters' parity agenda, dismissed them as extremist farmers who did not know what they wanted. Those activists who were attracted to protest-style politics, both within Grass Roots AAM and other organizations, participated with zeal and emotionalism bred on counterestablishment beliefs. Many protest-style farm leaders increasingly relished their fringe challenger status as radicals, tended to dismiss nonsupportive data on farm conditions as conspiratorial propaganda, and welcomed broad philosophical interpretations of social and economic conditions. In providing supportive information for the Farm Policy Reform Act proposal of 1985, these strategists were willing to admit that they selectively used data, took factual information out of context, and tried to mislead policymakers. In doing so, they were content to flout what their Washington-based activist colleagues and many peers in the reform movement knew to be lobbying convention. Those who wanted to portray themselves as "responsible representatives of family farmers" and who worked with agricultural specialists on economic analyses were, to policymakers, lost in the crowd.

Other protest leaders, especially those who assumed leadership positions in state organizations from the upper Midwest, perceived their fringe status differently. In particular, they wanted to break from the early AAM mold of threats, misinformation, and disruption. Their alternative has been to ally these groups with what farm leaders see as other exploited segments of American society, especially organized labor and racial minorities. Their efforts to include farmers within a broad populist coalition, as evidenced by support for Jesse Jackson's Rainbow Coalition and Senator Tom Harkin's (D-Iowa) New Populist Forum, have also linked churches, foundations, universities, and other institutions to the farm movement.

These activists did not worry about disregarding the conventions until after their identity was sealed in the eyes of most policymakers. Their antiestablishment attitudes, centered around a general belief in a victimized agriculture, successfully attracted a wide range of supporters long before farmers made overtures to social reformers. Both socialists and right-wing extremists felt their political times had arrived and became involved, especially in the smaller and more dissident groups. The anticorporate arguments served nearly every participant's political purpose. Most groups, especially at the state level, developed a set of neopopulist strategies for mobilizing popular support as a basis for both organizing and lobbying. These groups, as was the case with the AAM initially, were conceived of as alternative power bases for doing political battle with the enemies of agriculture. As they and many supporters saw it, the neopopulist farmers were, in the words of one activist, "just continuing the peoples' fight" against the many forms of corporate business and their macroeconomic strategies of "cost-price squeezing and the foisting of unmanageable debt on the family farmers."

Other farmers with more radically conservative beliefs were also served by the more extreme conspiratorial aspects of what the AAM's allies had rearticulated in rural America. Rightist activists attracted to farm rhetoric and rallies in which international bankers were denounced for conspiring to control markets and property conjured up elaborate economic theories. Through control of the world's major financial and industrial corporations, this "money power" was portrayed as establishing a comprehensive network of debt peonage. By manipulating currency and credit, the international bankers created inflation or depression at will. Workers, farmers, and independent businessmen were first made dependent on credit and then subjected to usurious banking practices and economic dislocations that caused them to default. Ensuing cycles of economic dislocation concentrated wealth in the hands of the multinational corporations controlled by international bankers.

The radical right organizations that sprang up throughout rural areas were diametrically opposed to neopopulist strategy because government was seen as too corrupted to be part of any solution to farm problems. Power politics as a direct struggle for governmental control was thus something to fear and loathe. Collective action was viewed as un-American and, often, un-Christian. As a result, radical rightists became militant individualists. Their proclaimed aim is to destroy centers of concentrated power and restore a pattern of social organization in which the individual, not the collectivity, is most important. Dispersed political power and maximum individual freedom from those who govern and lend money are their announced goals.

From 1982 through 1983, as new farmer assistance services and cooperative strategies were being formed, neopopulist and rightist activists shared some organizational unity. When the Iowa Farm Unity Coalition and Prairiefire organized local units, for example, they depended on a few radical right, or reactionary, sympathizers to fill out their ranks. In light of the leftist identification of the principal state organizers, this philosophical mix eventually made for some interorganizational tensions. As organizational activity and farm crisis rhetoric increased, so did factional problems. When central staff leaders felt compelled to speak out even moderately against the violence and threats advocated by rightist protest activists, their comments were met by threatening phone calls. Similar experiences of cooperation were noted in other states.

Other supporters, especially since 1983, have come from nonfarm ranks as both farmers and social activists have looked to a broadly based neopopulist coalition strategy at the grassroots level. The AAM has always received some farm-related business backing, especially in local communities, but the churches have had a much more sustained commitment to the agrarian groups in the past few years.

Church activists, in fact, have become instrumental in maintaining the neopopulist protest groups. Iowa Farm Unity, for instance, strongly urged that one local minister sit in with local organizers in each county unit. A minister, David Ostendorf, heads Prairiefire. Bishop Maurice J. Dingman of Des Moines, a past president of the National Catholic Rural Life Conference, serves as a forceful leader on midwestern farm issues. Several denominations employ agriculture, food, or rural specialists on their national staffs. All of these individuals have become involved with the farm protest groups. In addition, these churches support the National Council of Churches and its work with food and agricultural problems. Some church-affiliated activists who responded to the research for this study reported that religious groups supplied as much as 70 percent of the total organizational costs for farm protest activity spent in some midwestern states from 1983 to 1985.

The ideology and grassroots activism of protest organizations provides them with a solidary camaraderie. Obviously, religious leaders feel a need to identify with the problems of farmers from their own parishes and congregations. For many clerics, the neopopulist farm movement is also an extension of other social-justice causes that involved them earlier. Civil rights, urban poverty, the war in Vietnam, the environment, and world hunger seem closely related to the problems of the farm economy. A clearly articulated set of ideological forces, one of good farmers and one of bad corporations, helps bridge any perceived gaps between clerics and sons and daughters of the soil. These same premises, of course, are useful to evangelical ministers of the radical right who hope to win converts, save souls, and redirect moral values in a spiritually pure rural America.

Though new to the agrarian protest movement, the churches have been involved in the problems of their farm congregations since the depression of the 1930s. Arguing that ownership should be distributed widely to serve the greatest possible numbers,[21] church leaders have challenged the consolidation of privately owned farm land. Some church activists link current farm problems to developing the power base essential for the eventual enactment of redistributive land reform policies much like those in Latin America. The second important issue for the churches, and one their leaders see as closely related to ownership, is land stewardship, which has been given various interpretations ranging from chemical-free agriculture to soil conservation. The common position, however, assumes that large-scale agriculture cannot vest ownership in those who care for the land or its resources. Activist church leaders tend to believe in the inherent value of modes of production other than those capital-intensive ones to which agriculture has been moving for decades. Consequently, clerical supporters are greatly attracted to the agrarian ideals expressed by farm protesters.

Both through direct contacts and by publicizing the plight of farmers, churches have helped bring another nonfarm set of activists to agrarian protest. Since 1985, organizer ranks of the neopopulist groups have been swelled by a variety of nonclerics who were once social justice advocates for other causes. For example, the Nebraska Nuclear Weapons Freeze Campaign, Nebraskans for Peace, and No MX launched a campaign centered on "The Farm Crisis and the Arms Race." Minnesota Citizens Organizations Acting Together (COACT) brought all types of experienced activists and issue groups together to develop a cohesive nonviolent strategy of sit-ins and demonstrations to halt farm foreclosures. Ohio's well-established consumer action citizens' group, Public Interest, provided an office home and support to the floundering Family Farm Movement.

Without a protest-style ideology, mutual support between farmers and activists would never have come about. Issues of private ownership, government support of business enterprises, and the farmers' own history of support for property consolidation would have separated the farm activists from what many of them still refer to as "the hippy element." Anti-establishment attitudes, combined with a somewhat mythical interpretation of family farming as a glamorized alternative life-style, form the only bridge that leads to a shared organizational interest between the two groups.

Both solidarity and suspicion were evident at the National Farm and Food Conference held in January 1986 in Minneapolis. Cooperation was visible as participating activists directed panels to the farm crisis. Peace groups sponsored a panel on competition between farm and military programs for scarce tax dollars. Other nonfarm activists made use of their expertise in a variety of presentations.

The conference was the cooperative high point of interaction between farm and nonfarm protest activists, but the contradictions that made cooperation difficult were obvious even as the diverse participants endorsed one another.

Despite differences, the farm crisis as an issue of social justice was attractive to nonfarm conference participants. When asked about reasons for their participation, nonfarm activists made it clear that they considered the farm problem the only national issue of consequence that presently received both national attention and public support. That attention made the farmers' cause an expanding one. For various nonfarmer participants that cause provided philosophical solidarity, the excitement of a new protest movement, financial opportunities, and a forum for broadening support.

The continued attraction of a broadening base of supporters has made an ideology of power-directed protest a natural by-product of grassroots activism, even though ideology has become a counterproductive factor in formulating specific policy goals. Merging the farm agenda with those of peace, civil rights, urban, and world hunger activists heightens the problem of policy cooperation.

Radical rightists have contributed to this disarray. Despite their general shunning of power politics, many radical right farmers were attracted to the LaRouche movement as both supporters and candidates. When Lyndon LaRouche and his followers began to recruit slates of candidates for the 1986 election, they found it helpful to merge profarmer rhetoric with their strong criticisms of the poor along with their antidrug, anticommunist platform. Allied with a large mass of similarly alienated nonfarmers, farm spokespersons for LaRouche hit hard on conspiracy theory and violations of individual rights through U.S. agricultural and banking policies. Rightist LaRouche candidates charged established interests with concerted attempts to hold back commodity prices, bankrupt farmers, slash production, and conspire to starve the hungry residents of the Third World. These kinds of charges and related calls for reform further blurred many of the distinctions between neopopulists and the right.

These problems of beliefs and wants have created a distinct style for the agrarian protest groups. The grassroots groups as well as the AAM find it difficult to offer specific public policy positions. By broadening their support on the basis of anti-establishment rhetoric, and by refusing to accept bargained policy victories that are not directed at halting the exodus of farm families from agriculture, protest groups have grown apart from most other agricultural policy participants, especially after having proclaimed themselves the victims of "cheap food policy." At best, such groups can only keep calling attention to the worst aspects of the farm crisis and ally themselves with policymakers who are trying to cultivate personalized constituencies. These groups are still viewed as too extreme and unyielding to be acceptable policymaker partners. They refuse to tie the fate of the farm movement to policy participants who they fear will negotiate away higher farm prices. As a result, the agrarian reform groups, by adopting a unique representational style, are unavoidably confrontational as they escalate their opposition to national government and use more counterestablishment grassroots resources to press their case.[22] Because of the way established agricultural participants see the organization as part of a national movement of extremists, even an organization such as the AAM, Inc., with its belief in working with policymakers, will have an impossible time breaking from this cycle of escalating opposition. The only alternative either the AAM, Inc., or the National Save the Farm Coalition has to continued confrontational opposition to lower-priced commodity policy is to become a nationally important interest group. This choice, however, is unpalatable for many supporters because it means recognizing the reformists' resource limitations, restructuring their political goals, quieting demands, and attaching themselves to the political aspirations of someone who is a partner to the policy process. By doing so, however, agrarian reform leaders might contribute far less to the control of agricultural policy than the AAM did in the past, since reform representatives would be unable to force any of their own policy solutions before the public, the media, or even the remaining collection of agrarian protesters.

MEDIA SWEETHEARTS:
POLICY IMPLICATIONS

Without the media, the American Agriculture Movement could not have spread its message of farm problems, agrarian protest, and commitment to rural values. Because the organizers emphasized speed and efficient use of scarce resources, the AAM succeeded in mobilizing. In this case, the media made the message but only because AAM strategists knew that only television, radio, and press coverage could reach both farmers and the public. Considerable irony exists in the way a single organization used both a rancorous ideology premised on a traditional agrarian life-style and modern technology. AAM Chairman Marvin Meek summarized the AAM's public relations astuteness by concluding, "We may be stupid, but at least we're smart enough not to buy TV time."[23]

After the AAM began to cultivate their attention, media representatives followed the agricultural protesters. Initially reporters were attracted by the uniqueness of crowds of angry farmers who were able to mobilize so quickly in 1977. Throwing eggs at the secretary of agriculture only heightened that interest. Holding local parades, conducting a national tractorcade, releasing goats on the grounds of the Capitol, and assaulting a West German consul intensified coverage.

Gimmickry eventually gave way to serious concern for farm problems in attracting media attention, even after the disastrous 1979 tractorcade cost AAM protesters' support. By gaining and holding media coverage, AAM representatives and later those from other organizations were able to turn reporters' attention to the intense personal suffering of financially troubled farmers. Farm foreclosure sales provided the most dramatic opportunities to depict individual pain. Public relations consultants were attracted to the movement because they felt that farm conditions had gained the status of legitimate media events. From that point, however, violence needed to be replaced by human tragedies with which the audience could identify. Coverage of foreclosures and farm suicides succeeded in bringing reformers even more public support. By early 1986, 83 percent of respondents in a national CBS News poll felt that half or more of farmers were in financial trouble. In addition, while only 18 percent blamed government, 50 percent felt additional public financial support for farmers was necessary.[24]

The agrarian protest groups did not need consulting firms to help them make their point, even though they used them later in Washington to enhance their in-town political relations. Organizers and activists simply won friends and admirers among the press corps, a group of individuals inclined to view themselves as critics of government and exposers of wrong.[25] Hard work and the sincerity of the reformists' approach paid off in extra coverage. For example, the *Phil Donahue Show,* following up on a 1983 program with AAM activists, did a two-day telecast from Cedar Rapids in March 1986. The Iowa Farm Unity Coalition, operating in conjunction with the Donahue producers, brought a reported 10,000 farmers and farm supporters together for the show. At nearly the same time Dan Rather toured the farm belt for CBS News. This coverage was nothing more than an extension of the media attention that followed the protest movement after television networks first decided that the 1977 AAM meeting in Pueblo, Colorado, was newsworthy.

For the nonfarmer media and the attentive public, farm protesters have become *the* agricultural lobby. There are several reasons why. Actual policy differences, especially for something as complex as agriculture, are not easily argued through before a minimally attentive public. Time and space do not allow such coverage. Other farm and agribusiness lobbies, with vastly different views, have not been able to compete with the drama of the farm crisis, the poignancy of personal tragedies, the threat to entire life-styles, the attraction of vigils and protest events, or the personal appeal of protest group representatives. By the end of the 1985 farm bill debates, the widespread impression existed that grassroots activists ruled the climate of popular and farm opinion on the need for agricultural policy reform. This belief was articulated most clearly in Republican congressional fears about the outcome of the 1986 elections. Popular entertainers further reinforced popular sentiment. The farm crisis was covered in movies; and, in the case of *Country,* Prairiefire officials served as technical advisers. Through contacts established in film making, legislative spokespersons and farm activists arranged for three academy-award-winning actresses who had portrayed farm wives to testify before both the House Democratic Caucus Task Force on Agriculture and, courtesy of the television networks' nightly news, the entire nation. After their appearance, Senator Ted Kennedy included the statements of Jane Fonda, Jessica

Lange, Sissy Spacek, and the absent but supportive Sally Field in the May 6, 1985, *Congressional Record.* On September 22, 1985, in the midst of farm bill negotiations in Congress, over fifty performers, mostly country and western, staged the first Farm Aid concert as a farmer assistance benefit modeled after the Live Aid concert for African relief. AAM and farm crisis committee members dominated the planning, and one performer endorsed the Family Farm Reform Act to the roar of the crowd.

Monopoly status has a rather high price, however. Generating and sustaining a media campaign has taken still more attention away from serious consideration of how this array of fragmented interest groups can implement their very general political agenda. The media, group activists complain, nearly blitz state offices for help and assistance. Not all media attention advances organizational goals favorably when reporters focus on human rather than political elements of the crisis. "They ask for a farmer who will cry on camera," said one irritated staff person. An Iowa AAM official was asked to talk about personal feelings of suicide and murder rather than what she considered important farm problems.

The additional threats imposed on policymakers by media attention have also proven costly. Publicly acknowledged support for the reformers is often only reluctantly provided; and deep-seated personal animosity to protest activists exists among many legislators, bureaucrats, White House officials, and lobbyists from other groups. Unless the agrarian protest groups are able to work with midwestern state governments to create change, it may be that their greatest successes in reaching the public serve eventually to further isolate and undermine them. In a policy sense, the hope of one congressman may prove true. He envisions that "hangers-on will move on. Then everyone else will see that these guys have an empty cup and want a free ride." The entrenched ideology of policymaker opposition, the use of minimally substantiated data, and the uncertainty of continued press coverage make it unlikely that these groups will continue to win much support from agricultural policy participants if they continue in their diffusely focused opposition style. So does the likelihood that these zealous activists will just wear out in the face of continued farm losses. After all, the farm population suffered an 11.5 percent decrease from 1980 through 1985, with more losses projected.

Regardless of how many participants leave or wear out, or how much hostility these activists engender from policymakers, their residual effects will linger within the agricultural lobby. Too many farmers have learned to distrust their long-standing organizational affiliates. They have acquired considerable organizational skills, changed their attitudes about farm policy, and won some new allies in the process. These changes are too substantial to be negated soon.

5

Producers in Washington

Agricultural lobbying may be at a crossroads where two paths are so sharply divergent that each represents a fundamentally distinct alternative to policymaking. Lobbyists acknowledge that constituents' concerns have come to dominate Congress. Legislators who chart their own electoral courses with independent campaign committees and personal control over fund solicitation, media efforts, and issue choices have a great deal of freedom from the constraints not only of political party but also of any single interest.[1]

One variable that farm group representatives claim to worry about is legislative concern for constituent votes.[2] Congressional candidates, lobbyists fear, will choose among competing interests on the basis of how legislator support translates into votes. With this fear in mind, most farm groups rely heavily on constituents to plead the organization's case. Those who come from back home on a group's behalf bring with them a sense of what district voters want, or so legislators believe.

The crossroads that confronts Washington farm lobbyists has to do with the interests to which legislators will listen. Because much of agricultural policy is widely acknowledged to be a congressional prerogative in which farm interests are granted the greatest legitimacy, decisions about direction have to do with which farmers will gain recognition. Will it be the traditional lobby of general farm and commodity groups that continues to gain attention on the basis of the essential information they provide about program needs? Or will Congress turn to the farm protest groups that, despite their lack of popularity with policymakers, mobilize large numbers of potential voters because of their concern with farm financial conditions? With extensive public support for farms in crisis, a legislator's choice may be easy.

Impressions about constituents do not vary among farm state legislators and their staffs who discussed farm organizations for this study. Farm protesters appear sincere, financially distressed, and concerned. Farmer constituents recruited to Washington by established producer organizations often seem to be handpicked and unrepresentative. Moreover, state legislators—including the entire 105-member South Dakota legislature—and state directors of agriculture have arrived in Washington to speak for distressed farmers rather than for the 70 percent who are surviving.[3] Rural businesspeople, chambers of commerce, and implement dealers have also supported farmers with whom they have shared economic difficulties. Trade associations that oppose farm commodity programs have lost credibility when deserted by so many members. "Who do I listen to?" asked one congressional staffer. "I never see that happy son of a bitch from the Farm Bureau who supports Reagan who I keep hearing about. I only see the guys with the [protest group] caps."

The question, To whom Congress will listen? has important implications for policy. In an internationally competitive agriculture, decisions about production and pricing, once made, affect economic conditions for many years to come. Even though they might differ in specific approach, established farm groups believe that U.S. farmers must once again become internationally competitive after losing markets steadily since 1981 to expanding lower-priced foreign producers. Since 1983 their policy proposals have tended to reflect a need to expand U.S. exports and strengthen trade. Most farm protest leaders, on the other hand, fortified by an anti-internationalist ideology, disagree with such policies. Becoming competitive means lowering commodity prices in the face of an already unprofitable situation. An export enhancement approach with lower food prices would only hasten the exodus of many debt-plagued farmers from agriculture. Thus, to make farming profitable, demands for parity have been expressed in terms of restricted production, production controls, higher domestic prices, and import restrictions. Those protesters who do not favor isolation and systems of national food self-sufficiency, as articulated by many of their peers, add to their demands proposals for lower priced, or two-tiered, dumping of U.S. products on international markets.

If legislators maximize votes by turning to protest group demands, excluding established farm interests, adjustments will need to be made in the size and scale of U.S. agriculture. Avoiding these sharp measures has been a major preoccupation of farm group politics in the 1980s. Understanding this behavior is an instructive lesson in the meaning of interest group influence.

The farm lobby, as this chapter makes clear, is not an integrated establishment contending with a loosely structured opposition over the appropriate direction for agricultural policy. Recently formed and newly activated farm organizations have proliferated in national politics, each distinctly concerned with its own agenda of intensively focused policy demands. These new farm interests share little with older farm groups except their commitment to promoting programs that assign benefits to producers by commodity. This commodity emphasis, sharpened over the last three decades, has meant other important changes in the farm lobby as many commodity groups have enhanced their organizational capacity to represent a wide range of commodity-specific policy problems. It has also led to a lack of unity and purpose among these interests that neither the general farm groups nor anyone else can do much to restore. As a consequence, the traditional producer lobby in Washington—no matter how well organized—is no longer capable of dealing effectively with every issue that comes forward in a changing agriculture.

WHAT IS THE PRODUCER LOBBY?

The producer—or farm—lobby means different things to various respondents. It can be seen as a closely knit cadre of regular participants who know one another and who see Washington as essentially "a small town where everyone works in close confines without any secrets." The lobbyist who made this remark, and the dozens of policy participants who described national agricultural politics in similar fashion, was referring only to those organizations that mount a continuing lobbying effort.

The producer groups that always advertise their Washington presence include the Farm Bureau, National Farmers Union, National Farmers Organization, National Council of Farmer Cooperatives, and a handful of commodity interests. Corn, soybeans, wheat, cattle, dairy, sugar, cotton, tobacco, rice, and peanuts are the actively represented commodity groups seen in Washington as "the real players" on farm issues. With the exception of cattle and soybeans, all are subsidized crops in that their producers have been the prime beneficiaries of farm bill programs.[4] These commodity groups include both multipurpose and single-issue types. More importantly, from the perspective of the group representative, lobbyists from these interests are the ones who "get out and around." Their organizations allow these individuals the latitude to practice what one veteran lobbyist called "the art and science of lobbying." If there were an organized congressional Third House Regular, as associations of full-time lobbyists are sometimes called in state legislatures, representatives of these groups are the ones their peers would unhesitatingly admit.

Seen somewhat differently, there are "so many farm groups in town, really reputable ones, that no one knows what they all want. The number just grows and grows." The lobbyist respondent who made that remark was not referring to protest groups or to the producer/middleman organizations that less actively represent farm operatives. He and many other observers see more than just a handful of farm groups effectively represented in Washington. These other producer organizations are believed to have a decided, though not continuing, policy impact.

For various reasons, ranging from the size of the commodity harvest to the issues that bring the groups to Washington, these organizations fail to enter the highly interactive flow of lobbyist traffic regularly. Ideals of a Third House Regular or Washington-as-small-town do the representatives of many such organizations a decided injustice by relegating them to a second-class status. These organizations, rather than being portrayed as deviates from some norm, must be seen as typical Washington interests.

As was noted in Chapter 2, most agricultural organizations are involved in the ebb and flow of politics. Only a small percentage of major groups are afforded the luxury of paying daily attention to their place in the agricultural arena. That does not preclude other interests—often with the help of consultants—from organizing, developing access to policymakers, fostering an analytic capacity, supplying useful information, and, in the process, providing important services to their producer members. A greater general emphasis on constituent linkages in lobbying makes periodic, single-project style, political forays all the more feasible for even isolated regional interests.

These are the producer organizations that Washingtonians often dislike for the uncertainty and the surprises they may bring to policymaking. Because their organized presence is not routine, many policy participants find them irritants who "muck up the place, make it difficult to gain and hold an agreement." Such organizations add the precise dimension to group politics that Jack Walker and his research associates found responsible for impeding bargaining and destabilizing policy partnerships throughout national government.[5] In effect, because so many of these organizations come and go from politics, the nonregular farm lobbies continually threaten to overload the policy process with more information about producer needs than can be handled effectively. Several kinds of conditions are likely to activate these interests. Intra-organizational conflict, domestic policy changes, and shifting international trade relations account for much of this behavior.

INTRA-ORGANIZATIONAL CONFLICT

Numerous examples of conflict within groups have led to both the formation of new organizations and the activation of existing groups. The National Farmers Union was formed in 1902 from factions of the National Farmers' Alliance. Fifty years later, the U.S. Farmers Association (USFA) split from the NFU.[6] Although the Farmers Union was the most liberal farm group of its day, Iowa Farmers Union activists with former Farm Holiday ties went their own way and formed the USFA because its leaders believed that the national organization failed to address key social issues of the 1950s. The Farmers Association never became anything more than a midwestern regional group, but President Fred Stover kept reappearing in Washington to speak for civil rights and peace causes whenever he believed that the NFU was negligent. When the Farmers Union failed to back the parity demands of the newly formed American Agriculture Movement, the USFA attempted to fill that void. In this way, the U.S. Farmers Association kept coming back to national politics whenever tensions heightened between organizational parent and offspring.

While such predictable lobbying cycles are hard to identify among farm groups, respondents see the pattern of organizational parents begetting national political activist groups as often repeated. An example of cyclic involvement is evident whenever the American Farm Bureau Federation refuses to support the tobacco program. State farm bureaus from the tobacco states both threaten the national and get actively involved on their own in Washington. This behavior has become thoroughly expected. A less expected example of noncyclic Washington activism occurred when the Alabama Farm Bureau withdrew from the AFBF over what dissidents saw as the federation's lack of response to the farm crisis.[7] Unlike other state farm bureaus, the Alabamans went on to hire their own Washington representatives who would become active on support issues where the Alabamans and American Farm Bureau differed.

Commodity and cooperative groups, their lobbyists believe, are not immune from such conflicts simply because their interests are defined somewhat more narrowly than those of the general farm groups. In 1985, the Land O'Lakes co-op broke with the National Milk Producers Federation over dairy assistance programs. Regional cooperatives, affected differently by milk surpluses, exhibited sharp disagreements over diversion and herd buyout programs. Representatives of the financially plagued Land O'Lakes felt it necessary to break ranks and lobby against some Milk Producer initiatives. The Farm Credit Council initially faced less open opposition from some of its less financially hard-pressed cooperative Farm Credit Districts that feared that some proposals for the 1985 Farm Credit Bill would direct their financial resources elsewhere within the farm credit system. Their representatives had few fears about suddenly stepping in, however, even litigating against the final legislation. Several of the twelve districts had never sent representatives to Washington previously, either independently or to work with the council.

Wheat producers within the National Association of Wheat Growers (NAWG) were split even more publicly in Washington than were dairy producers. Several state organizations, most notably in Nebraska, were active proponents of farm protest movement goals. Dissidents were especially angry with the Wheat Growers' willingness to accept lower loan rates in hopes of lower international grain prices. Though many wheat farmers who were members of the NAWG worked actively with the AAM and the Nebraska Farm Crisis Committee, the national was able to hold state delegations together only by allowing them considerable latitude. The Nebraska Wheat Board went on to hire consultants to dramatize farm problems, using Gray and Company to conduct a public relations effort and Chase Econometrics to do a financial analysis. Because of its ability to push the national, several key legislators viewed the Nebraska Wheat Board as an independent and significant voice in Washington.

Other organizations appear on the national scene less because of conflict over policies than because of differences as to *which* policies should be addressed. The National Cattlemen's Association (NCA) lobbied as a typical single-issue organization until late in the 1970s, concerned mostly with protecting subsidized grazing on public lands. Cattlemen lobbyists, much to many members' dismay, seldom addressed issues that did not directly influence cattle and calf prices. Other organizations, at NCA member requests, began to lobby on selected issues in response. The Texas Cattlefeeders Association (TCA) hired a Washington attorney lobbyist to actively work on tax policy. Soon afterward, the newly formed Independent Cattlemen's Association began to lobby for import quotas and brucellosis inspection, two issues that NCA was accused of lagging behind on because of a reluctance to offend Department of Agriculture friends. By 1985, with pressure from competing organizations, the National Cattlemen had a much larger lobbyist staff and had become a multipurpose organization that carefully followed these and other issues. Both the TCA and the ICA moved out of the active policy arena into watchdog involvement.

DOMESTIC POLICY CHANGES

David Truman regarded changing policy conditions as responsible for a political disequilibrium that, in turn, activated group responses.[7] This was presumed to be responsible for the wave phenomenon noted earlier. While Truman may have tried to explain too much about group behavior on the basis of organizational disruptions, policy changes and threats of change certainly bring already organized groups to Washington.

The Farm Bureau and its state affiliates are a particularly good example, and one that demonstrates the great utility found in the AFBF's extensive organizational structure that allows representation from many sources on nearly any issue. Ten major commodities are well represented by fifteen or so organizations that employ the regularly active commodity lobbyists, but those commodities represent only a portion of U.S. agricultural production. California, the most diversified agricultural state, produces over two hundred distinct products. Michigan, the next most diversified state, produces over fifty crops, well over half of which are not in competition with those from California. While proportionally few of either state's products are directly subsidized, most producers in California and Michigan are frequently affected by national issues related to farm labor, water, and nutrition. Marketing order legislation is also of prime concern. Price support programs create indirect effects on many smaller commodities and become relevant policies. For example, diverted-acreage provisions in the 1985 farm bill allowed program participants to plant unlimited amounts of nonprogram crops in place of such supported products as corn and soybeans. Competition could have been devastating to many farmers if surpluses of vegetables and leguminous crops had been allowed to flood the market, forcing prices lower.

Since the Farm Bureau's Washington office divides its ten lobbyists according to specialized policy concerns, state bureaus frequently step in to represent crops that only their lobbyists know well. When an issue affects independent citrus growers, for example, the California Farm Bureau may well be more active in Washington than the producer/middleman-organized United Fresh Fruit and Vegetable Association.

By no means does the need for this periodic intervention fall exclusively to national organizations and their constituent groups. The alliance of co-ops, the Agricultural Council of California, frequently intervenes; but so do smaller California grower organizations. California kiwi growers organized in 1979. Many of the nearly fifty other fruit, vegetable, and specialty crop growers worked directly with the Agricultural Marketing Service to develop the orders that support grade and size standards, standardization of packaging, market flow regulations, market allocations, reserve pools, producer allotments, research and development, and advertising.[8] When policy questions arise or order changes are proposed, representatives of these organizations typically use their own lobbyists. Conflict over marketing order restrictions on navel oranges, for example, was played out in national politics mainly by two state interests in 1984 through 1986. A single grower, Carl Pescosolido, mounted an intense campaign to challenge the order, initially received food marketing industry support, and was met by opposition led by Sunkist Growers. The issue was so specific and so controversial to many members of other producer groups that Sunkist Growers was left as the main, albeit formidable, defendant. Sunkist's defense was so thorough that the cooperative was even able to get attention from both the media and the office of the secretary of agriculture in undermining the credibility of a report from the Economic Research Service that questioned grower benefits from the navel orange order. In subjecting the ERS report to careful scrutiny, Sunkist analysts were able to capitalize on the failure of the ERS model to consider adequately the effects of weather.

The many issues and the laws that govern them are often so complex that such things as marketing orders are not handled by regular farm lobbyists. Water issues in the valleys of California and guest worker provisions for alien farm laborers are not only of interest to a few growers; they are also complicated issues to negotiate. As a consequence, the affected interests simply feel more comfortable intervening themselves. When a farm bill provision such as acreage diversion creates negative effects for other commodities, as so frequently happens when a move to produce market equilibrium in one commodity leads to disequilibrium in another, such an error is often one of policymaking omission. Mistakes such as this seldom fail to cause a reaction. In studying the Food Security Act of 1985, the Michigan Bean Council was the first to recognize and act on the implications of diversion for nonprogram crop growers. Normally not active in Washington or involved in farm bill politics, the Michigan group leaders immediately contacted their congressman, Bill Schuette, a first-term Republican on the House Agriculture Committee. Schuette and a second Michigan legislator, Democrat Robert Traxler, worked with the Bean Council to sponsor corrective legislation as soon as Congress convened in 1986. By March, after other nonprogram crop groups and farmers who produce everything from guar to platago ovato had joined the Dry Bean coalition, the Schuette bill passed, and much of the problem was alleviated. The American Dairy Association (ADA), the nonlobby marketing arm of the dairy industry, is another example of specialized intervention independent of more traditional organizations. In 1986 the ADA waged a campaign against the restrictive milk-advertising policy of the Federal Trade Commission. Prompted by legitimate marketing responsibilities, that action was taken in conjunction with public relations support from yet another group, the National Dairy Promotion and Research Board.

Major issues and commodity programs also attract minor organizations into the policy process, often with what policymakers see as a significant impact. Some of these organizations are trying to correct problems as the Michigan Bean Council did in 1986. Farmers for Fairness, a new group, began to work on inequities created by peanut program provisions of the 1981 farm bill that had passed on quota cutbacks disproportionately to small tenant farmers. After failing legislatively, Farmers for Fairness and Capitol Hill supporters held an extensive series of protests directed toward changes in administrative interpretations of the present program in both 1981 and 1982. In 1986, the North Carolina Grange was able to develop a plan that permitted the tobacco program to pass. Tobacco strategists, fearful of consumer and health opposition in a general farm bill, had opted to include that program in the general tax reform bill rather than in the Food Security Act. Not only did tax reform fail, but there was no agreement over its tobacco provisions in 1985. Only the expertise of the Grange, working with Senator Jesse Helms (R-N.C.), produced a tobacco compromise that passed in a small deficit-reduction bill that also included state oil and gas exploration revenue and medicare reimbursements.

At other times, smaller organizations contribute more of a cosmetic policy effect. The Association of Family Farmers, an organization of family farmers of unknown origins, makes political contacts with congressional tax committees to gain special tax advantages. Other financial interests find association support useful as part of a broad defensive strategy on tax policy. Dairy Farmers for Responsible Dairy Policy, a group of large-scale producers, played the same role for a coalition of interests opposed to dairy programs in the 1985 farm bill. By serving as a farm organization front, opponents hoped that dairy farmers could add further credibility to the opposition. In neither case were these organizations intended to be the kind of central policy players Farmers for Fairness or the North Carolina Grange turned out to be.

INTERNATIONAL TRADE RELATIONS

Except for being somewhat more volatile, trade policy creates extensions of group politics similar to domestic issues. For example, the National Pork Producers Council, a nominally involved low-key free trade proponent, suddenly began to support the imposition of trade quotas on Canada. Of course, the Pork Producers had also claimed to be supporters of President Reagan's commodity program initiatives on the 1985 farm bill. Members, especially Iowa Pork Producers, promptly forgot that more general concern after developing a farm bill provision to include a $1 pork checkoff plan to be imposed on hog producers for enhancing marketing and advertising efforts. The Pork Producers are only one example of general organizational support bending to the immediacy of new political issues.

The volatility of this issue area has much to do with problems of U.S. farm exports. As policymakers sought to increase production in the face of escalating worldwide demand, United States food policy was reversed with the Agriculture and Consumer Protection Act of 1973. Instead of curtailing production for most major program crops, farmers were encouraged to plant fence row to fence row.[9] However, problems for producers began almost immediately. By 1974, price instability was becoming evident as international demand fluctuated by commodity with changes in the economics and politics of buyer nations. Demand increased until 1981, however, when commodity exports began a series of annual declines that were much in evidence as the 1985 farm bill was being drafted.

What this meant for producer groups was a shift in what their representatives see as the general philosophies that had grown up around many of these organizations, especially the larger ones. On international issues, that philosophy often took a free-market or open-trade approach. Sugar, cotton, and dairy interests, along with the NFU and the NFO, were long-time advocates of import restrictions and quotas, but most of the other farm lobbies that addressed the issue were philosophically free traders. Those principles had developed as practical responses to import restrictions on U.S. products long before U.S. agricultural supremacy was challenged in the international market. Organizations voiced free-trade beliefs in opposition to such countries as Japan, which restricts imports of food from the United States. In addition, the Wheat Growers had not envisioned the European Economic Community (EEC) as a net exporter of that commodity; the Cattlemen could not envision a competitive Argentinian beef industry; nor did the American Soybean Association see South American countries growing and then exporting large amounts of their product. Nor were American rice producers and the Rice Millers' Federation inclined to see major gains in agricultural development turning many Southeast Asian countries into rice exporters instead of customers.

Developmental gains and technical innovations in shipping agricultural products to the United States have led many of the most active farm lobbies to scrutinize individual trade issues and agricultural policies within competitor nations. As early as 1982, when export declines were just beginning, the Farm Bureau was sending its own delegation to the European Economic Community to complain about the EEC's subsidy policy. Such involvement is now an expected part of trade policy negotiations. Sometimes farm groups are useful for U.S. administration trade efforts—for example, when leather interest complaints in 1985 led to Japanese compensatory trade concessions worth $260 million. At other times, group complaints are particularly unwelcome, as they were when U.S. vegetable growers provoked a controversy over competitive Mexican agricultural development just south of the Rio Grande. The State Department had endorsed that development as an export-generating device for Mexico's troubled economy.

The involvement of less frequent farm group players further discourages government-wide U.S. policies to encourage open trade and combat the subsidized mercantilist trade policies of other nations, especially as these generally remained unchallenged during most of the Reagan administration. Among the traditional producer groups, the National Wool Growers Association frequently involves its scarce resources in policies. In 1985, Wool Grower representatives actively supported the quota proposals in the Textiles and Apparels Trade Bill. While failing there, the Wool Growers Association was able to gain a 20 cent tariff increase on New Zealand lamb.

Other producer groups, usually not identified as lobbies, are more frequent activists on trade policy than are many of the smaller traditional organizations. These are the more than 100 groups supported by state agricultural checkoff programs whose funds go to domestic and international marketing efforts through such nationally organized associations as the U.S. Feed Grains Council, National Peanut Council, and Tobacco Associates. Nearly 70 of these state and national marketing organizations belong to the U.S. Agricultural Export Development Council and participate in developing trade policies and strategies. Sixty of them also belong to the program of the Foreign Agricultural Service in which they share market development costs. Just by participating and offering trade proposals, not to mention making demands on administrators, these organizations help shape much of U.S. food trade policy over such important foreign-aid matters as technology transfer and economic assistance.

While many of the checkoff groups limit their political involvement to cooperator programs, the representatives of other groups believe that they use their positions and political contacts to good advantage even when organizational bylaws prohibit direct lobbying. Some lobby through other organizations to which they have close ties and overlapping membership. The U.S. Feed Grains Council, for example, works through the Corn Growers. Others simply lobby directly with a low-profile approach to disseminating information. In 1985, cooperator organizations were involved extensively in negotiating export subsidies and cargo preference for U.S. shippers and export subsidies. As organizations directly in contact with foreign governments, these groups also develop policy positions on the basis of maximizing trade advantages for the interests they represent. To some extent, such proposals are often advanced with the association serving as stand-in representatives for foreign agriculture. With declining U.S. exports blamed on foreign-subsidy programs, attacks on various countries are equally likely, however.

The political conditions that attract less active groups to politics are numerous, and they demonstrate how dynamic an agricultural issue can be in attracting just farmer—or producer—involvement. Given the diversity of farm products and farm economics, these groups play a useful role in filling representational voids in agriculture. Any view of the farm lobby that fails to acknowledge the role these less active interests play is surely inadequate.

ISSUE INVOLVEMENT

The large number of farm groups that demonstrate their policy interest and the extensive array of issues that activate them provide evidence of the pluralism that characterizes the entire agricultural lobby. Farmers, as organized in Washington, are not a monolithic force. Yet, even while emphasizing the comprehensive attention that farm groups give to agricultural issues, a distinct bias must be acknowledged both in terms of what most concerns farm organizations and what direction this concern takes. Since this bias is directed, rather unidimensionally, toward commodity price support programs while many groups are moving to diversify their representational involvement and their capacity to lobby effectively, farm interests must continue to be defined as much by paradox as anything else.[10]

Representatives of nearly all the established producer groups believe that their organizations must be involved in an extensive number of issues that take them beyond the boundaries of traditional agricultural policymaking. The farm groups cope in many ways. The most obvious method is to reshape the organization's lobbying structure into a multipurpose one, much as the general farm organizations have always done. The Farm Bureau is still considered the best-organized multipurpose model. With a total staff of twenty-eight, the Farm Bureau employs an executive director for the Washington office and ten other public affairs specialists, each of whom lobbies both agencies and Capitol Hill. Responsibilities are divided by subject, with one technical support specialist and nine lobbyists responsible for: transportation; farm policy, price supports, and field crops; labor; agriculture-environment; marketing, nutrition and livestock; tax, budget, and appropriations; energy, water, and public lands; credit, insurance, and social issues; and international trade. The remaining Washington staff provides ancilliary services, including research and analysis. The AFBF's office meets the needs for addressing a changing policy environment.

The Farm Bureau, however, had only to adapt its previous functional style to a contemporary emphasis on the total business needs of a modern agriculture. Whereas twenty-five years ago four Farm Bureau lobbyists addressed less complex farm issues as well as many conservative political issues of both economic and social consequence, present lobbyists find it difficult to cover all farm business problems. Social issues and nonfarm economic problems, though often still included in national Farm Bureau resolutions, can hardly be addressed if AFBF lobbyists hope to cover farm issues adequately and maintain reputations based on pragmatism and expertise.

Financial and personal resources do not allow any of the other farm groups to replicate the Farm Bureau's comprehensive structure, even as it would be modified to a different set of concerns. The other general farm organizations and farmer-owned cooperatives, with the exception of the well-staffed National Council of Farmer Cooperatives, employ one to three full-time lobbyists for agricultural issues. These lobbyists must keep priorities in perspective while remaining able to address whatever problems arise within the organization.

The commodity organizations that have a multipurpose structure are better staffed to handle those responsibilities than are most of the generalist organizations. During the 1980s the National Cattlemen's Association grew from a staff of six to twelve, becoming the most comprehensively organized commodity organization. The National Milk Producers Federation employs four lobbyists, as do the National Wheat Growers and the National Cotton Council. Each of these organizations also uses professional assistance from their members, and they have several in-house support personnel. The similarly staffed National Broiler Council and United Fresh Fruit and Vegetable Association also represent multiple interests on behalf of their industries.

Given the complex responsibilities of these six commodity organizations, their lobbying operations must be extensive. For example, the Cattlemen represent both feeder and cow/calf interests. Even on an issue such as farm credit, this distinction becomes important. Cattle feeders use mostly short-term credit, while cow/calf operations need long-term loans to develop desirable herd characteristics over several generations. Hence both tax and credit issues vary by producer type. Once an organization follows a multipurpose strategy, the range of issues often continues to broaden. Nutrition issues, for instance, serve as a logical extension of group efforts to deal with both disease and marketing problems. As NCA members experienced consumer problems with such health warnings as those about cholesterol intake, the group made plans to add a representative to work with the National Institute of Health and similar public health agencies.

Quite unexpectedly, according to nearly all their lobbyists, a multipurpose commodity group can find itself working on several issues at once, no matter how hard it tries to set priorities. Many of these issues and the participating policymakers are less familiar to group representatives than are those with which they are involved more frequently. For example, during the last quarter of 1985, with the farm bill in jeopardy, the Wheat Growers found themselves juggling several issues. The group was considering the applicability of the marketing loan concept, arguing against production controls, attempting to maintain target price supports for wheat, aiding grain reserve proponents, working for a cargo preference compromise, attending to pesticide regulations, and fighting tax reform. The last two activities were not even farm bill provisions.

Most producer organizations are not prepared to take on so many issues. Yet their members are no less troubled by problems throughout the policy process. Group strategies to deal with this burden, all of which focus on retaining their identities as single-issue or project groups, vary. The common denominator in coping with so many issues while attending to a few involves heavy reliance on other organizations. The National Turkey Federation, whose members are affected by many of the same regulations as the broiler industry, relies extensively on the Broiler Council to work out USDA safety and inspection requirements. The Broiler Council also negotiates regulatory compromises on behalf of all poultry producers with the Food and Drug Administration, the Environmental Protection Agency, and the Occupational Safety and Health Administration. The Turkey Federation also usually follows and endorses Broiler Council leads on legislative proposals. The National Cattlemen occupy the same leadership position for general meat industry issues. Because of Cattlemen efforts, the Wool Growers and Pork Producers can have smaller Washington offices and not be too disadvantaged. The National Corn Growers, in the same way, rely on the Wheat Growers.

These informal arrangements have what lobbyists see as additional advantages. The lead groups benefit from the nominal support and endorsement of the less involved organizations. Coalitions can be formed quite easily when knowledgeable and experienced lobbyists can be brought in on an issue and assist Cattlemen or Broiler Council staffs. Wheat Grower lobbyists cite this as a reason why they were happy to see the Corn Growers open a full-time Washington office. Potentially threatening policy disagreements over broad issues can be minimized by a long-standing sense of cooperation.

Finally, by concentrating on just a few issues, groups such as the Wool Growers and Corn Growers are able to operate effective Washington lobbies. The Wool Growers' single lobbyist, freed from attending to other matters, was able to direct his attention to five critical sheep and wool issues in 1985: reaffirming the Wool Act as part of the farm bill, increasing the New Zealand lamb tariff, transferring the Animal Damage Control Program to USDA, gaining a noncut classification in the Gramm-Rudman-Hollings deficit reduction bill, and supporting the textiles trade bill. No other groups allied themselves with the Wool Growers on three of the first four issues, all of which were victories. The Corn Growers concentrated only on farm credit, corn price support, and ethanol issues in 1985, ignoring pleas to become active on pesticide regulations and tax reform. While acknowledging that they felt lost when a member inquiry did direct them to the Environmental Protection Agency or the Commerce Department, most of the single-issue-style commodity lobbyists considered their groups just as successful on core product issues as any of the better-staffed groups.

A negative side of single-issue involvement, in addition to ignoring some important issues, is the dependent status the group and its lobbyists may come to occupy on some policy areas. Lobbyists from the Wool Growers, Corn Growers, American Sugar Beet Growers, the American Beekeeping Federation, and any other groups that must restrict their active participation find their own unique interests and demands merged with those of other organizations, coalitions, or individual policymakers. To a great extent, except for a few especially close policymakers, these single-issue groups lose much of their identity. Should their policy needs or problems involve them in extended political discussions, these organizations can easily find themselves without any independence in bargaining situations because other policy participants neither understand their political positions nor recognize their political resources. This was evident in the 1985 farm bill when congressional conference committee members quickly stripped away previously hard-won sunflower producer benefits while restoring those of honey producers. Honey interests had a well-established policymaking partner in conference who allowed them to win back gains lost in the agricultural committees. Sunflower interests did not.

The single-project commodity group that occasionally comes to Washington is even more precariously placed than are those normally in Washington. Regional sugar, peanut, and tobacco groups find it necessary to use consultants and industry representatives, such as the Tobacco Institute, to interpret the policy maze and gain a recognizable policy agreement, even on major commodity provisions. Home state representatives, no matter how favorably disposed to constituents, have insufficient knowledge and influence to put together an agreement among several regional interests. Regional fragmentation of economic interests led to great problems for the sugar program in 1974, the peanut program in 1981, and the tobacco program in 1985 and 1986. When represented with careful attention and detail, however, single-project groups are successful even in the face of consumer or public-health opposition. On issues of less immediate concern to members than the commodity price support programs—such as agricultural research or pesticide regulations—these single-project interests have been heavily dependent on and supportive of state farm bureaus and other general farm groups.

The irony of this situation is that the more a farm group benefits from its capacity to address many issues, the less it matters for the organization's primary interests. While the Wheat Growers can be effective in resolving divisive issues, such as cargo preference shipping provisions, they are no more likely to get what they want for price support levels than are either the Corn Growers or the coalesced Peanut Growers Group. Moreover, the Farm Bureau, history tells, is less likely to win on these issues than are narrower commodity interests, even though the AFBF and the other general farm groups give first priority to such price support matters.[11]

Commodity price support programs have not taken second or third priority to other policy decisions. Despite the new prominence of other issues and despite lobbyists who want to work first on policies they can win, commodity programs come first for nearly all producer groups. Why? Because, at least in the minds of those who manage farm organizations, producers get involved directly as group activists—even though they may not affiliate for this reason—for monetary gain. The members whom staff personnel talk to, lobbyists claim, expect farm groups to represent economic interests rather than conservation, consumers, or the world's poor. In Anthony Downs's applicable terminology, commodity programs are the heartland issues for producer organizations.[12] As tools for determining profit and loss, price support programs demand the greatest maintenance and honing. Since they are up for renewal in each farm bill, they are subject to annual congressional modification to reflect operational problems or farm economic conditions, and are adjusted routinely by the secretary of agriculture under statutory provisions of the farm bill. For example, the secretary is given a range of loan levels under farm bill provisions that need be set and later adjusted at the discretion of the administration.

Organizations, according to respondents, use three sets of cues about what producer members want: farm economic conditions, group activists' views, and current knowledge about how programs are contributing to recipient income. These cues come in the broad context of farmers' reactions to government and public policy. Sometimes the cues lead to modification of commodity programs. At other times, they bring about a full-scale attack on the programs or those who make them. Given farmers' awareness and the constant fluctuations in commodity prices, organizations cannot assume the latitude to ignore heartland concerns. Nor do these cues lead to an orchestrated and comprehensive approach to the financial problems of all farmers.

This is the point at which producer groups find their distinctive policy niche if they choose to develop a political identity. On basic commodity price support programs, the commodity organizations can mount an all-out defense of their members' economic interest. For each active organization, this means independently addressing the wool program, corn program, wheat program, dairy program, peanut program, honey program, and so on. This defense usually requires considerable political adjustment, as it did in 1985 when commodity organizations—looking at product surpluses, international trade conditions, and political circumstances—developed strategies to keep members' incomes as high as possible while dealing with ways to eliminate product oversupply. For feed grain interests, this meant accepting considerably lower loan rates in the hope that declining U.S. prices for these commodities would lead to increased exports. In return for lowering the loan rate, these organizations gained continuing support levels for the commodity target prices that trigger deficiency payments to producers. Thus, when the lower loan rate for wheat produced a lower commodity price, farmers who sold at that price gained increased deficiency payments.

The Milk Producers reacted similarly even though dairy farmers receive no direct support payments. After facing two major legislative acts, which severed price supports from parity and instituted voluntary supply-management in the dairy program, dairy interests knew that they had to reduce production to save their program in 1985. The co-ops fought for and won provisions for whole herd buyouts and direct diversion payments as voluntary cutback measures. In return, dairy farmers were to receive minimally lower prices for dairy products purchased by government under the dairy price support program.

None of this adjustment in program benefits comes easily; producer members prefer maximal benefits, which their lobbyists do not find realistic. As a consequence, farmer members are themselves farmed hard to gain support for a program proposal that has a chance of passing. Regional peanut producers who formed the Peanut Growers Group after suffering costly losses in the 1977 and 1981 farm bills wanted to recoup those losses in 1985. Considerable food industry opposition to the peanut program and cost factor concerns with the total farm bill package made recovery unlikely. However, any concerted effort to restore gains, in the views of peanut lobbyists, could have taken so long that opponents would have steamrollered the entire program and its system of price-enhancing restrictive quotas. "Timing is everything," explained one representative. But convincing financially hard-pressed growers of that point is never easy.

Despite the disclaimers of several organizations that claim to want less of what even their representatives call "special interest" legislation, there is little immunity to the attractions of commodity programs. The American Soybean Association, an organization whose staff members are fond of calling themselves "free traders in theory and practice," opposes direct price supports as mercantilist policy and was willing to go along with lower 1986–1990 loan rates for soybeans to promote exports. However, they first insisted upon direct compensatory payments to cover soybean producer losses. The meat and poultry industries, without commodity programs of their own, carefully scrutinize what goes on in other programs that may affect their producers. Dairy herd buyouts, with resulting increases in beef supplies, caused the Cattlemen to lobby hard for more government beef purchases to offset member losses. Feed grain prices, as affected by commodity programs, are always an issue for organizations such as the Cattlemen and the Broiler Council, which represent the multiple interests of large users who are necessarily cost-conscious.

For other commodity interests, marketing orders take the place of direct payments, loans, and quotas as their first priority commodity programs. Many of these organizations have no other policy concerns except for trade issues. For even a major organization such as Sunkist Growers, protecting the navel orange marketing order takes economic precedence over such issues as pesticide safety and employment regulations for farm workers. The order determines the price and, thus, the profit. Even when marketing orders do not invoke restrictions on quantity, producers hope that orderly marketing provisions will stabilize product flow and provide better price returns.[13]

This fragmentation of interests, centered around an extremely large number of independent programs, does nothing to facilitate a cooperative and interdependent farm lobby in Washington. There is no umbrella organization to broker major issues that separate the groups, as equitable public funding does for dairy. The commodity organizations do their part to make agricultural policy piecemeal. In looking toward a comprehensive farm support policy, the organizations of cooperatives are in no position to serve as brokers or intermediaries. The cooperative associations accord the commodity programs a priority status because they must assist rather than lead agriculture while attending to nonagricultural issues. The old Cooperative League of the USA, now the National Cooperatives Business Association, was long noted for support of its co-op members' dairy programs. The National Rural Electric Cooperatives Association understands that rural electrification issues depend on farm group goodwill, since farmer support will not be forthcoming on the basis of the co-ops' generally high rates.[14] The National Council of Farmer Cooperatives provides important support for marketing orders and the nonpricing provisions of commodity programs for farm producer cooperatives, but its members' competing interests leave it in no position to negotiate levels of commodity supports.

This leaves the willing but no longer able general farm organizations to broker agreements over commodity programs. Although the NFU and the NFO have been supportive of high levels of price supports for commodity programs, the two groups combined are neither sufficiently staffed nor trusted to negotiate intercommodity agreements. In fact, the NFU and the NFO rank and file are seen as too willing to adopt policy positions inconsistent with U.S. export needs; many commodity representatives identify both the NFU and the NFO as overinvolved with the agrarian protest groups.

The Farm Bureau is different. Many of the most actively represented commodity organizations, especially from the Midwest, have been willing to come together under the auspices of the AFBF to discuss mutual problems that might jeopardize their programs. The AFBF has been eager to broker price provisions, discuss program costs, and since 1981, move from its historically conservative position on support levels. As a consequence, the Farm Bureau has gained what representatives of other groups see as a tenuous acceptance as a colleague in commodity program policy. Meetings in 1983, 1984, and 1985 produced important points of agreement that were followed, for example, in setting the lower loan rates for the 1985 farm bill. However, even the most supportive commodity lobbyists see the Farm Bureau in a service role in which AFBF representatives help determine such important policy parameters as the total attainable cost for commodity programs. No one, except perhaps some of its own officials, sees the Farm Bureau as a power broker that can force agreement over individual program costs or concepts and drag dissenting commodity interests along. The commodity groups, and those they hire to represent them, have been too successful on their own over the past three decades to think that an interloper is necessary to stabilize programs or pass commodity legislation. "But that doesn't mean I don't want Farm Bureau to represent [us]," concluded one commodity lobbyist, "I do. But I want those guys getting involved with agricultural chemicals, research, genetics, that stuff. Those are important issues that even [we] don't have time for."

POLICY IMPLICATIONS
OF A FRAGMENTED FARM LOBBY

Were it not for the high-impact presence of grassroots farm organizations, the producer part of the agricultural lobby would have to be described as decentralized. The split between the many groups that want to negotiate on their own in Washington and the array of only semicooperative groups that want policy reform changes that description to fragmented.

As noted, the Washington farm lobby is an extensive one, composed of both regular participants and a myriad of occasional activists. Though the farm lobby is organized around an extensive and growing list of issues and policies, the emphasis of most farm groups is directed toward income maintenance commodity programs. Only a few farm organizations have the structural capacity even to begin to attend to the wide range of problems facing a modern agriculture. There are, as a result, many agricultural concerns that farm representatives pay far less attention to than many agricultural observers might have guessed. Attention to farm groups is limited by other priorities, the expressed urgency of farmers and co-op officials, lobbyist's perceptions of the consequences of addressing a problem, and the imminence of the issue.

These limitations mean not only that the farm lobby in Washington is unlikely to be the watchdog for a rapidly changing agriculture but that it also is the adjustment vehicle for demanding short-term policy tools for producers threatened by disallocations in the market. For example, farm groups are content in the mid-1980s to deal with the inherent price instability of export markets as long as public policies exist to smooth over anticipated financial rough spots. Given the problems of oversupply and surplus for major crops, farm lobbyists have seen no alternatives that would be palatable to their members, even though they have expressed widespread discontent with farm policy.[15] Though central to a few marketing orders, production controls have few adherents among farm lobbyists, especially because large-scale farmer members dislike those policy tools and are reportedly the most vocal activists within these groups. For the same reason, targeting of income supports to some percentage of producers most in need, rather than through generally received commodity programs, has had no open advocates among Washington farm groups. While farming has undergone an organizational and management transformation, its representatives have given only marginal attention to those specific increases that account for this structural change: size, concentration of ownership, use of capital goods, carrying risk, and credit.[16] Moreover, of all those farm groups, only the Farm Bureau, Cattlemen, and National Council of Farmer Cooperatives (with its organizational spin-off Farm Credit Council) have the staff and resources to assume issue leadership in analyzing the policy implications of these changes.

There is, therefore, no clear indicator that the established farm groups are better prepared than the grassroots reformers to provide policymaking assistance in sorting through the answers to questions about the future direction of a changing U.S. agriculture.[17] While farm organizations have demonstrated appropriate analytical skills, these skills have continued to be applied to basic commodity programs, both for price supports and for other income assistance. Agrarian protesters, while sometimes discrediting themselves in their policy approach, nonetheless have asked questions about problems of size, absentee ownership, land stewardship, and alternative agriculture. In addressing changing conditions, they have demonstrated to many congressional policymakers that they may have a policy relevance, even though they continue to express most of their demands in the form of better-funded commodity programs rather than on the basis of comprehensive reform.

For policymakers looking for guidance on how best to cope with a changing agriculture, the choices can easily appear to be between established policy participants who provide status quo support for somewhat more market-oriented commodity programs and farmer activists who want at least to address the meaning of agricultural change. Two additional factors must be considered. First, the traditional farm lobby has been seen to be at least somewhat splintered by organizations that have come to Washington with policy goals shaped by protest group sentiments. The Nebraska Wheat Board and Alabama Farm Bureau are only two examples. The reluctant support of the NFU and the NFO for a producer referendum over mandatory production controls in the 1985 farm bill reinforced feelings about the strong influence of grassroots farmers. Farm bill participants shared a widely held belief that only a verbal battering of NFU and NFO leaders by grassroots activists won their endorsement of production control amendments. Second, as noted before, throughout the 1980s protesting farmers have remained more visible than have supporters of established farm interests. Policymakers who have been under sustained protest group pressure find it difficult not to believe that these groups most accurately represent the opinions of farmers and rural votes.

If legislative candidates have to make definitive choices between these two sets of farm interests in the mid-1980s at the height of the farm crisis, the balance of representational power might well swing to the generalist demands of agrarian protest. Absolute choices seldom need to be made on policy decisions, however. The years 1980–1986, especially the later ones, were years of accommodation. For example, none of the more active farm groups could argue for or accept immediate limitations in direct producer support on commodity programs. Lower loans rates did not translate into many actual dollar losses for farmers. Even the more conservative Farm Bureau uncharacteristically advocated high levels of direct supports despite Reagan administration complaints and AFBF concerns over high budget deficits. The wisdom of AFBF's conciliatory approach was evident when Farm Bureau delegates, in early 1986, elected Iowa Farm Bureau President Dean Kleckner to head the national organization. Kleckner's Iowa organization had cooperated to at least some degree with Iowa Farm Unity, Prairiefire, and the state chapter of the United Auto Workers on state farm problems.

Because of policymakers' expressed concerns over farm protest, traditional farm interests simply escalated their financial demands on behalf of existing commodity programs. Legislators, otherwise uncertain about how best to satisfy protester demands for more farm income, had little choice but to go along. The end result for the Food Security Act of 1985 was that the two sets of farm interests shared credit for a farm bill that looked much like the one organizations on both sides had earlier rejected. This bifurcated influence was acknowledged most precisely in the widely circulated comment from a USDA source who stated that the protest movement added $10 billion dollars in costs to the initially estimated $52 billion three year (1986-1988) commodity program package in the bill.

Even if policy choices between competing farm interests became more distinct, should organizations such as the National Save the Family Farm Coalition succeed in formulating specific policy alternatives, there would probably still be no easily determined victories. Some reconciliation would surely continue as organizations fought to maintain their claims to influence. More importantly, choices about which farm interests best represent agricultural policy are not made in a political vacuum. Agribusiness, consumers, and others would surely intervene as their organizational representatives sought to protect and defend these interests.

6

Agribusiness:
Lobbying and Adjustment

Farm group rhetoric often accuses agribusiness of suppressing producers through both pricing mechanisms and public policy initiatives.[1] During the 1980s, agribusinesses have been singled out by farm group representatives for considerable blame for the troubled farm economy: the Chicago Board of Trade manipulates farm prices; grain companies profit from farmers' losses and then, abroad, sell wheat contaminated with dirt; processors and manufacturers will do anything to hold prices down. By 1984, farm organizations were denouncing an agribusiness plot to support the Reagan administration on the 1985 farm bill.

This chapter finds little support for that contention, however, especially in the context of the 1980s. The agribusiness lobby is nearly as divided as the Washington farm lobby, both in its preferences about the direction of agricultural policy and in its intensive focus on narrow issues. Except for the greater reluctance of most agribusinesses to mount an aggressive lobbying effort on farm policy, there are few differences in the organization and lobbying problems of the two often presumed opponents. Agribusiness, however, is unique in the long-range focus of its public affairs work, a characteristic that lends subtlety and some subterfuge to much of its lobbying effort. Nonetheless, the lobbying priorities of agribusiness complement those of established farm groups as often as they compete with them.

As demonstrated by the complexities of the food system, farm interests would be expected to find that the public policy positions of agribusiness are not always compatible with those of producers. In addition, events of the past fifteen years have helped draw a policy distinction between farm and business interests.

The 1973 farm bill marked an important departure in the willingness of food retailers to accept the outcomes of whatever farm legislation producer groups help
negotiate. Consumer boycotts, primarily of meats, directed the greatest ill will to
grocers, not farmers. Retail interests, with marketing strategies at stake and
plenty of alternative products available to promote, began to study what public
policy decisions they should support.[2] Other factors have also been at work. Concentration within grocery manufacturing and food retail industries continue to
reinforce business's willingness to take the consumer's side. Mergers have increased the buying and promotional power of both manufacturers and retailers,
often at the expense of producers' traditional market shares.[3] Ten years later, the
1983 Payment-in-Kind (PIK) program and its reduced plantings rallied many
other producer industries around a similar resolve to scrutinize policy proposals
and resist economically distasteful ones. To an important extent, agribusinesses
have changed; they no longer condone the idea that farm legislation is a legitimate producer prerogative. In that sense, they are much like the rest of the business community.[4]

Differences between farm and agribusiness interests have not escalated
into political warfare between these two forces, however. Farmers and businesses share a desire for stable markets and production without the damaging
financial effects of severe price swings. Production and marketing arrangements have been coordinated for many years to minimize farm income losses
from producer prices that were too low and commodity costs that were too high
for middlemen.[5] Commodity exchange contracts have been used profitably to
hedge against market swings, for example. Early-season contracting between
buyers and producers also stabilizes financial conditions over the course of several months for both parties. Of course, many producers have moved into processing and distribution. In short, there are many commonalities that tie
various interests together within the food and fiber systems, even in the face of
increasingly pluralistic policy involvement.

Many of the complaints farm group leaders register against agribusiness reflect little more than anticorporate rhetoric—speaking to a captive audience on a
favorite subject and finding a target for blame. Criticism from farm groups provides the additional strategic advantage of further dividing agribusiness interests, with whom producers are indeed competing for shares in returns on the
consumer's food dollar.

Agribusiness is no less diverse than production agriculture. Given the conditions industries face in an international marketplace, leaders can hardly share a
common vision of an ideal agricultural policy. Despite concentration and merger,
sometimes on an international scale, differences in self-interest continue between those who hope to exchange value-added products at a profit. The way
business leaders deal with these differences varies, too, of course. Sometimes agribusiness interests pursue public policy demands aggressively through active
lobbying. At other times, other business interests simply adjust their operations

to government decisions and work toward profit via different avenues, resulting in actions and relationships that bring agribusiness interests together with farm organizations, both cooperatively and as opponents, often in unpredictable ways that do not reflect the heated rhetoric between the two.

A POLITICAL RETICENCE?

Given the verbosity and visibility of producer interests, a close observer of agricultural politics would be struck initially by agribusiness's lack of open public-policy involvement. There are several reasons why. First, agribusinesses become involved in comparatively fewer agricultural policy issues than do farm groups. Second, they are involved in a variety of food and trade issues far removed from basic farm concerns. Third, the agribusiness lobby tends to be both more long-term in its approach and more responsive to industry image problems than does the established farm lobby. Fourth, agribusinesses accomplish much of their public-affairs work through a variety of trade associations. Finally, though farm organizations believe that their policy demands are best served by playing for public support, agribusiness interests prefer unobtrusive lobbyists and would rather arrange a quiet one-on-one meeting between a legislator and a district businessperson. This combination of reasons makes it appear as if the mass-membership farm organization operated as a different type of lobby than the industrial agriculture lobby. In reality, the differences are more a matter of strategy and participant relations within the two lobbies than any big differences of policymaking style. As Lewis Anthony Dexter implied in his study of organizational representation in Washington, strategies and contacts reflect representational problems and needs.[6] The kind of interest involved is only one factor affecting problems of representation.

Less Involved

Though it would be tempting to conclude that agribusiness interests are usually more selective about agricultural issues than are farm interests, that interpretation would be inaccurate. A larger percentage of agricultural issues are potentially of a nonadjustable higher cost to producers than they are to business. The National Cattlemen, for instance, see several public programs directly affecting their members' economic standing. For cattle feeders, brucellosis inspection, meat imports, government beef purchases, whole herd dairy buyouts, animal safety, health and nutrition studies, and feedgrains prices will have economic consequences that may not necessarily be passed on to consumers. Cattlemen lobbyists cannot ignore government plans and those of other private interests by simply adjusting to the consequences of disease, competition, and rising input costs and then raising beef prices above a cost of production. Despite its customers' affection for beef, the Food Marketing Institute feels no such pressure to think of these items as major heartland issues. Most of their economic self-interest on these issues can be realized through a hard and fast support of free trade. Beyond that, a service-oriented FMI can help its retailer members adjust by assisting in the merchandising and promotion of alternative products such as seafood. As long as profits are maintained, grocers care little if consumers buy beef or poultry, apples or pears, cane sugar or corn sweeteners. Since the food industry, through market research, views 50 percent of its consumers as flexible in their food habits, these services are important.

This variation in the comparative issue involvement of producer and agribusiness interests reflects the different type of diversification characteristics of the two sets of organizations. Although most producer groups—including those commodities that have taken on middleman functions—are divided by product, separating one crop from another within the farm sector, another quite different diversification separates agribusinesses. Agribusiness organizations divide by functional sector—manufacturing, wholesaling, retailing—within the overall dynamics of the food or fiber system. When producers care about more than one part of the food system, that concern is due to growing more than one crop and supporting more than one commodity organization's policy agenda. Agribusinesses, in contrast, frequently diversify vertically and across the system by commodity as products undergo change. Cargill, for example, not only manages nonagricultural businesses, it also merchandises grain, processes meat, mills corn and wheat, refines corn sweeteners and various oils, produces seed, and manufactures animal feed. Though Cargill may have more industrial problems, producer lobbies more frequently deal with government on behalf of those issues that affect their single product.

The historical governance of agriculture on a crop-by-crop basis, with the resulting proliferation of groups in Washington, accounts for this frenzy of producer activism. Several policy decisions must therefore be made for different products, whereas a single resolution can often satisfy the remainder of agricultural interests. For example, when federal authorities set shipping rates on inland waterways, those rates are then in effect, even though they may vary by vessel and type of goods shipped. Once loan rates have been set for the peanut program, however, only peanuts are covered, and then only for a limited time. Those involved with wheat, corn, and other crops still need to deal with the issue. Differences in production methods and susceptibility to disease and pests further complicate producer group problems since issues of chemical use, farm labor, and agricultural research also demand crop-by-crop attention. Agribusinesses do not face different food safety regulations just because they produce food snacks rather than breakfast cereals.

These differences in policy emphasis between producer and agribusiness lobbies can be seen in some of the lobbies' organizational features. Little splintering occurs between organizations that share a generic interest. State affiliates of agribusiness interests are common, but no policy differences were observed within the trade associations in national politics. Carefully defined marketing and lobbying responsibilities, in effect, limit the potential number of competing participants within agribusiness sectors. Producer interests exhibited no such stability in the 1980s. In the few instances in which seemingly competitive trade associations exist, the individual organizations do not represent distinct positions on issues of varying policy priorities. The two peanut product associations, American Peanut Products Manufacturers, Inc., and the older Peanut Butter and Nut Processors' Association, are separated mainly by personality and because certain members won the right to set internal organizational policy. The Independent Bakers Association and the more politically active American Bakers Association co-exist for the same reasons, as do the Food Marketing Institute and the National Grocers Association. A more viable and businesslike distinction can be seen in the frozen food industry; but even those political differences are nominal. The American Frozen Food Institute (AFFI) operates a multipurpose lobby with extensive policy interests and also manages affiliated organizations for frozen potatoes and pizzas. The much smaller National Frozen Food Association (NFFA), with a processor and manufacturing membership that overlaps somewhat with the AFFI, pursues legislation and administrative rulings that principally promote sales to government-supported institutional food users. Buyers are also members, and NFFA managers receive a commission on cases of products sold for providing material services.

The multipurpose organizational format, though narrowly directed toward certain types of businesses, is also more commonplace for agribusiness than it is for producer groups. Except for isolated examples such as the NFFA, single-project agribusiness associations that pursue their own financial goals on the basis of public policies that enhance their own contracts are rare. Other than the few small, single-issue organizations such as the National Independent Dairy-Foods Association and some of the politically less active foodstuffs associations, most of the trade groups provide a wide variety of selective member benefits. Services are especially broad for those organizations whose membership needs either directly or indirectly bring them into product merchandising. Of the more than fifty nationally organized food and fiber associations, over half are multipurpose. All of them lobby on a wide variety of agricultural issues, but nearly all spend far more staff time on product assistance and generic marketing and research than their public affairs representatives do on congressional and administrative matters.[7] The ideas that participants share in conferences, executive planning sessions, and technical seminars often set the public affairs agenda in a far more informal way than do those generated at executive board meetings and membership conventions of the farm groups. New products and new marketing strategies almost always bring the businesses into governmental regulatory decisions.

Most agribusinesses, both firms and trade associations, consider agricultural policy rather narrowly as commodity policy, even though they must act on many more issues. Other food and fiber issues are categorized according to their relevance to the specific organizations as a retail or manufacturing problems. Even a large firm or trade association typically assigns one or perhaps two public-affairs representatives to commodity and other farm policy. Smaller organizations give other responsibilities to their agricultural specialists, or they hire consultants as needed. An input industry may assign still other staff members responsibilities for research, international trade, energy, and tax policy. A typical output industry would also assign personnel to business policy, consumer issues, food standards, and nutrition. The consumer issues specialist for an organization such as the Grocery Manufacturers Association spends an inordinate amount of time with the Federal Trade Commission on advertising and labeling concerns and therefore directs more attention to the Food and Drug Administration and the Department of Commerce than to the USDA. The same is true on legislative matters and the appropriate congressional committees other than agriculture. While the multipurpose producer organizations surveyed in 1985 and 1986 each gave some priority status to between five and nine issues involving farm commodities and production, multipurpose agribusiness interests typically gave priority consideration to only one to three of these issues. Many organizations listed no policy priorities of this kind. Under these circumstances, agribusiness's collective reputation as an agricultural lobby can hardly be as well established as that of the producer interests that get involved more extensively. Only a few agribusiness organizations are exceptions to this rule, and then only because they pay attention to a single production issue such as seed legislation or a wheat reserve.

Other Issues

Agribusiness representatives repeatedly identified five issues and bills that consumed their time, both in preparation and lobbying, between 1983 and 1986: tax reform, the superfund for environmental clean-up, trade regulations, product lobbying, and the Federal Insecticide, Fungicide and Rodenticide Act (FIFRA). Provisions of the 1985 farm bill other than trade items were mentioned far less frequently, and, in over half of the instances in which the farm bill was mentioned, all of its provisions were considered to be of secondary importance to some combination of the other issues.

The responsiveness to these issues demonstrates the role that agribusiness plays within the broader context of agricultural—as opposed to just production or farm—policy. To an important degree, agribusinesses remain the primary interest group spokespersons for issues critical to food and international policy aspects of agriculture, except for the protectionist positions of several farm groups on trade issues and other specific commodity trade decisions. Only on FIFRA were producer groups as active as agribusiness and, even then, that bill did not receive priority status.

The five highest priority concerns, beyond simple industry profits, could be summed up in terms of market expansion or eliminating further market loss. Tax reform, as it appeared to be shaped through executive and legislative compromise in 1985, was viewed negatively as costly to product innovation and supplies of many commodities. Since both of these factors are important in marketing alternative products that either substitute for other commodities or can be used for their novel appeal to non-price-conscious consumers, the food industry was felt to be especially susceptible to damage from tax reform. The Superfund issue, with a choice between a tax on oil and chemicals or on value-added products, offered the same consequences. Food industry and ethanol interests opposed the value-added tax, but with the cost situation reversed, agrichemical and high energy user interests favored it. In these two instances, producer organizations were getting representation that only the Cattlemen, Broiler Council, and Farm Bureau were addressing on an active basis from a farm perspective.

For similar reasons trade regulations were targeted as agribusiness issues. Although the specifics varied from one organization to another, the underlying difficulties agribusinesses address are the protectionist policies imposed by other nations and the perceived reluctance of State Department officials to combat them. As a result, several agribusiness interests have decided to work with the Foreign Agricultural Service, apart from the cooperators' program, on developing strategies for removing trade barriers. Other organizations have more restricted goals. The American Meat Institute, for example, needs United States-European reciprocity on packer quality standards to facilitate its members' entry into the European market. The Grocery Manufacturers of America and processor associations need to address the difficulties faced by their multinational membership in setting up European plants. Almost the entire public affairs effort to increase U.S. food product sales abroad through value-added exports has fallen to agribusiness.

Product lobbying and FIFRA have a different appeal to agribusiness. Part of the concern is with liability. That is, how extensive must warning labels for tobacco and other products be before the administration decides that the risk is no longer the producer's but the consumer's? If regulatory actions and harsh lobbying are too restrictive, product use will decline. Because of the importance of inorganic substances in minimizing disease and pest problems, restrictive use also translates into decreasing commodity supplies and increasing prices. Another concern troubles food industry representatives, however. Although plentiful supplies of inexpensive products are always necessary and desirable, another growth area for market expansion is quality, high-cost food. One dimension of food quality is associated with the wholesome image of chemical-free safety. To this extent, food industry buyers have urged cautious use of agricultural inputs for the purpose of generating consumer goodwill toward more expensive products. Large-scale manufacturers and grocers, in particular, want to avoid losing this portion of the market to small firms and health food stores. These competing agribusiness problems have made any issue of chemical use a complex policy decision that attracts multiple interests.

To most agribusinesses—except those that sell fewer units of farm input equipment and supplies when acreage or other reductions are applied—farm bill provisions are less important than these other issues. Provisions for production quotas, production controls, and diversion of producer efforts automatically generate opposition, since fertilizer and feed interests have no comparable alternatives to grocers and food manufacturers. As a result, these agribusiness input interests set a higher priority on farm bill provisions than do most user organizations that oppose dairy, sugar, or peanut programs.

The real concerns of commodity-user agribusinesses can be seen better in the provision of the 1985 farm bill to which they collectively assigned top priority: clear title. This risk-reducing provision essentially awarded clear ownership, or title, of commodities to producers rather than to lenders. In the event of sales to a processor or manufacturer, that ownership would be transferable and not subject to any lien or mortgage. Given the precarious financial condition of many producers in 1985, most food industry representatives considered a stable market flow more important than a small reduction in commodity prices or even a lowering of loan rates. Though most agribusinesses felt the latter goal to be important, their representatives based priorities on the premise that producer organizations would accomplish that objective on their own. Behavior of this sort further reduces agribusiness's reputation on matters of agricultural policy, even though its lobbyists almost exclusively represent all of agriculture on many other critical but less topical and identifiable issues of food and fiber use.

LONG-TERM STRATEGIES

A farm organization lobbyist accused agribusiness of being slow: "They're talking about the 1981 [farm] bill and planning for the one in 1990. But here we are in 1985 with a vote on our amendments coming up tomorrow. [The agribusiness lobbyists] that I'm working with just do not seem to realize that Congress can and does rework these damned bills every year." Most agribusiness representatives, not so burdened with the shifts of farm bill legislation as it involves the rewriting of commodity price support programs, indeed do not think solely about short-term issues. Most of the firms and corporate boards to which these lobbyists are responsible would neither like nor sanction it.

When fifty respondents were asked what determines an agribusiness firm's approach to public affairs, staff lobbyists and hired consultants unanimously identified the singular importance of the chief executive officer (CEO) and, sometimes, that official's closest advisers. According to those whose incomes are dependent on them, CEOs set the tone for the work of both individual corporate public affairs staffs and the staffs of closely monitored trade associations. Moreover, even more than producer counterparts, respondents perceived conspicuous differences concerning what lobbying entailed between headquarters and Washington representatives of nearly every agribusiness firm. Although many Washington lobbyists for agribusiness liked an independently aggressive, flexible, and comprehensive public affairs strategy, these employees also believed that most CEOs were suspicious of such endeavors. Why? Representatives thought that CEOs rose through the ranks with an insularity from Washington affairs and even politics in general. Isolation bred too much reliance on the conventional distrust and suspicions of politics and its delegates, even its own.[8]

Several organizational conditions result from these disparate beliefs about lobbying, according to the Washington respondents. First, a tension between headquarters officials and Washington staffs becomes almost a necessary byproduct of their relationships. Second, most lobbying staffs, of both associations and firms, are asked to rely heavily on the technical expertise of corporate employees or consultants chosen by the CEOs rather than on their own analysts. While the firm gains control, lobbyists lose much of their capacity to adjust quickly to the ever present daily shifts in sentiment among policymakers. Third, lobbyists lack the respect and status within most organizations to serve as effective brokers on intrafirm disagreements. Compromises over policy decisions are too often reached far from what active lobbyists see as the real world of politics. Fourth, lobbyists feel that they are most valued for the information they convey to headquarters rather than for what they can accomplish as advocates. Headquarters' directives imply—and sometimes directly state—that Washington agribusiness representatives should be low-key in their work with policymakers and that they avoid offending anyone. When a major issue or a crisis occurs, CEOs and their personal advisers either come to Washington and take charge or back away from the problem. In both situations, lobbyists feel that their own best opportunities to provide organizational services are lost.

Those respondents who perceived the negative consequences of tensions between CEOs and representatives shared a view of lobbying expressed by Lewis Anthony Dexter after his engagement in an extensive study of business in politics. Dexter emphasized that the Washington representative's job must be more that direct advocacy or mere surveillance.[9] The lobbyist can serve the needs of the firm by offering strategic advice and assistance to those in the organization who, for example, receive public attention. Public relations gains can be maximized and negative actions minimized. In Washington and in national affairs generally, lobbyists feel that they can be coordinators of a comprehensive public relations effort in which policymakers, the media, and potentially useful allies are cultivated. Discussions of national and industry needs, the tactical means to these ends, and politically important information can elicit favorable images of the firm's policy concerns that can be crucial to its long-term goals. Interorganizational tensions can make such contributions difficult.

At the worst, however, intrafirm tensions can lead to what many of agribusiness's employees see as "brutally ineffective public affairs." In evaluating their colleagues from other organizations, several of the most active representatives noted that the antipolitical attitudes of company officials often led to low pay for lobbyists, the hiring of inferior personnel, the use of Washington as a dumping ground for officials dismissed from corporate offices, lackadaisical attention to political events, and few incentives for attending to details.

These conditions do not negate agribusiness's collective influence. They do, however, explain agribusiness's limited policy involvement, as well as the general impression that, compared with producer organizations, its representatives fail to attend to those issues that an outside observer might expect of multipurpose organizations with wealthy financial backers. Given Dexter's much earlier findings, this limited involvement and the accompanying criticisms of its own lobbyists also suggest that most agribusinesses may not have mounted as extensive a lobbying effort as their counterparts in other industries. While agribusiness lobbyists spend considerable time putting out costly policy fires on tax reform and directly advocating the use of oil revenues for Superfund, headquarters personnel are often more comfortable with lobbying efforts that address such longer-range considerations as comprehensive trade policy. These kinds of well-developed policy positions are more in keeping with the planning and operations research models of industries. When its representatives decided to enter the policy debates over the 1977 farm bill, Cargill drafted a conceptual Farm Income Support Program as an alternative policy approach. A proactive trade association looks to potential policy problems over a period of several years and maintains a commitment to addressing them, even if immediate conditions fail to warrant such attention. For example, in 1985, in the face of low wheat prices and commodity surpluses, the American Bakers Association insisted on lobbying for several proposals protective of their members' desire for a stable, low-cost supply of wheat. The ABA not only chose to oppose mandatory production controls, but its lobbyist also worked for a wheat reserve and expressed concerns about the need to limit Soviet wheat sales. While the Fertilizer Institute battled to win on the 1985 farm bill, much of its effort went to positioning itself for future farm legislation as far away as 1990.

The long-range interests of agribusiness firms are also addressed through extensive public relations campaigns that, like forward looking policy positions, can be orchestrated by headquarters personnel. Many of these campaigns simply show the strengths of lagging individual products. The National Soybean Processors Association's "liquid gold" promotion advertised the portability, export importance, and nutritional value of soybean oil for Third World residents. Part of that approach was aimed at increased sales under the Food for Peace program, but it was also directed at fixing attention on the importance of refined or value-added product exports in general. The tobacco and cotton industries made very similar appeals directed at winning U.S. consumers to their products, in view of declining sales.

More importantly, public relations efforts are used, in the words of one lobbyist, to "correct widely held misperceptions," especially when these misperceptions create public policy problems for a firm or industry.[10] Cargill, the only grain-trading firm willing to reveal its operations before a congressional hearing on the industry, found it necessary to begin a public offensive on the firm's behalf after facing the grain-trading scandals of the mid-1970s.[11] Corporate officials testify before Congress and buttonhole legislators. Media contacts are used to encourage stories favorable to the firm. Advertising is used extensively on television, on radio, in the press, in magazines, and in trade journals.[12] Smaller-scale campaigns often result, not through a lack of effort by the initiators, but because agriculturally related problems are so frequently raised and of such generally limited popular appeal that the media give them only minimal coverage. For example, beef industry representatives had to settle for little more than an advertising campaign to direct attention to the nutritional value of their product in the face of medical warnings. Few press sources were willing to cover the dispute between cattle producers and nutritionists until the concept of "lite" beef, as a unique health-oriented response, gained some popularity.

Advocates of a more provocative and controversial proposal, food irradiation, had no trouble finding public officials and journalists who would discuss that issue at length. This attention provided the food preservation technique an extensive public forum. To obtain agribusiness's end, coverage from the media need not be directed to the specific issue or policy goals, however. Food industry representatives, concerned with promoting expensive processed products, were pleased with the degree to which they were able to interest journalists in the preferred eating habits of yuppies and baby boomers. This coverage both advertised the product and made policymakers aware of its economic significance. "It's important to demonstrate that [consumer] appeal," noted one processor/manufacturer lobbyist, "because frozen foods gain many policy advantages that can only be maintained if these products are considered essential to the food diet."

COMPANY DIFFERENCES IN STRATEGY

Not all firms or industries carry out their conventionally styled lobbying, even over long-term issues, in the same way. Several lobbyist respondents see four models of public affairs used by agribusiness companies that at least occasionally want to be partners in the policy process. When applied to production or farm issues of agriculture policy, specific firms stand out as most typical of each model.

Archer-Daniels-Midland represents the most aggressive of these lobbying models. Its CEO uses firm officials, a trusted lobbyist consultant with whom he has daily contact, and several other retained Washington representatives to give a "hard-sell" in promoting pending policy goals. Each participant has well-defined responsibilities, a set of assignments, and no fear of ADM's taking a controversial public position in Washington. ADM lobbies individual policymakers, does not hesitate to do favors for legislators, makes extensive use of electoral contributions through its owner's family, creates direct ties to the White House, and builds alliances with whatever organized interests may help serve the firm's goals. Its lobbyists work hard to win friends from both political parties and from executive and legislative leaders. ADM's political strategy encompasses whatever tactics will work. Its CEO sought to head an important presidential commission on international private enterprise to state the firm's views.[13] At the same time that the firm was winning friends, however, ADM was willing to make enemies by arranging for oil company officials to be called to the White House "woodshed," as one lobbyist described it, for a high-level executive brokering of ADM and oil business conflicts over public support for ethanol manufacturing and sales. Rather than relying on public-relations campaigns to promote corn sweeteners, ADM representatives have provided active support for the sugar programs as a means of making corn sweeteners more competitive economically. For the same reason, ADM fought to protect sugar quotas, even as its opponents argued that the United States must help support Caribbean initiatives aimed at that region's political stability. Lobbyists and consultants maintain that, on farm programs, no other agribusiness firm but ADM duplicates the assertive lobbying characteristic of farm interests.

Cargill, however, is nearly as prone to take the offensive as is ADM, and many lobbyists consider the firm the most influential agribusiness in Washington. Cargill officials, and those who follow this approach, do not work Capitol Hill or the White House as extensively as does ADM. Favors and elaborate use of campaign funds are avoided as potentially harmful to Cargill's image as an honest trader. But Cargill lobbyists, nonetheless, are well known in both executive and legislative offices for explaining well-defined corporate positions. Cargill executives, including the experienced public affairs director and veteran chief lobbyist, meet regularly to plan policy positions and strategies. Though these meetings can lead to specific proposals on farm price supports—as they did for the 1977 farm bill—Cargill generally addresses related issues such as grain quality and cargo preferences to union shippers. As part of its self-described soft-sell approach, Cargill places a heavy premium on both proactive and reactive public relations, preparing and disseminating policy statements, and advertising.[14] These techniques are usually sufficient to establish Cargill's free-market trade position. Rather than relishing an all-or-nothing win, Cargill representatives prefer to appear conciliatory. A typical Cargill position on an issue would state the firm's willingness, for example, to pay its fair share—but no more—on Mississippi River shipping rates. Allies who have worked with Cargill note that the executives bargain hard, even on their soft sells. If a compromise can be scaled on a basis of one to ten, these respondents note, Cargill will always go for a low-cost two as long as that goal seems attainable and not detrimental to the firm's public image. Cargill's image-conscious public relations approach was evident in early 1985 when its grain traders were able to buy red winter wheat from Argentina and ready to ship it to North Dakota for milling at a price lower than U.S. farmers could provide. The purchase received wide notoriety at a time when economists were critical of high U.S. commodity prices. When faced with farm protest and a potentially negative press, Cargill backed off and sold the wheat elsewhere at a loss. Nonetheless, Cargill made its point.

The third model is best represented by Monsanto Corporation, one of several agribusiness firms just beginning to think through the costs and benefits of being active on a wide variety of production issues. Over the last five to ten years, the executives of these firms have become convinced that agricultural programs other than agrichemical regulation affect their economic well-being. While Monsanto has long lobbied hard on a few selected agrichemical issues, much as ADM and the National Association of Wheat Growers do for their interests, its representatives have not addressed specific farm problems other than to observe quite generally that its officials favor free trade and a free-market agriculture. The question for firms such as Monsanto is how they should best become involved in agricultural issues, since their executives and representatives lack both credibility with policymakers and expertise. Answers have been worked through slowly and with careful attention to the details of each firm's public image. Monsanto's approach in the 1980s has been to "help elevate preliminary farm bill debates to a higher, more informed, level." The firm participates in policy conferences, helps sponsor research, brings in consultants to clarify positions for its executives, and, in preparation for the 1985 farm bill, conducted a series of regional and national Dialogue conferences on the conditions and policy needs of U.S. agriculture. Firms such as Monsanto have been willing to engage in direct lobbying only on such heavily opposed issues as farmer referendums on mandatory production controls.

The vast majority of agribusiness firms are involved less directly in varied agriculture issues than is Monsanto. Firms like Continental Grain prefer to do nearly all their position taking, lobbying, and public relations through the many trade associations to which they belong. The representatives of these firms are primarily watchdogs, studying policy positions of Congress, the administration, and other private interests. While experienced lobbyists like those of Continental have reputations for exercising considerable influence within the trade associations, they prefer not to work outside the associations, even on a soft-sell basis. If these firms' executives do not agree with trade association positions or if they are unable to block them, they usually do little more than temporarily distance themselves from association lobbying and wait to try again.

TRADE ASSOCIATIONS AND OTHER ORGANIZATIONS

Firms with multiple interests, vast numbers of potential issues to watch, and small Washington staffs use trade associations as an economical and efficient means of representation. For agribusinesses with executives favoring long-term solutions and positive images as a representational approach, trade associations also provide considerable anonymity. Safeway and Giantway supermarket chains, for example, find it preferable to have the Food Marketing Institute address controversial issues since the FMI insulates them from potential customer backlash. A typically large firm may belong to eight to twelve trade associations and related organizations.

Trade associations offer other policy benefits. They provide an obvious vehicle for a united front on any issue for which member support can be mobilized. Associations can also develop an expertise and familiarity with a narrower range of issues than can most firms. Diversification and mergers, not to mention the many facets of any single business, have brought most agribusiness firms to a point at which generic issues such as milling, foreign sales, or domestic distribution give way to what seems best for the conglomerate concerns of the company. Thus, trade associations often provide a useful counterperspective on pending problems and issues. Little else holds competitive industries together for any common purpose. Associations have a greater flexibility in developing working relationships with other interests than do individual firms. Personal difficulties, as well as legal restrictions, make it difficult for a single manufacturer to develop a working coalition with organized labor or suppliers of any one of its raw materials.

Despite these advantages, reliance on trade associations produces some negative features associated with the low visibility of agribusiness in the total dynamics of agricultural politics. Many firms that keep a low political profile lose their individual identities, even to some extent with their own congressional delegations. Only those firms that lobby directly, doing more than testifying before legislative hearings or attending fund raisers, were recognizable to most congressional staff personnel who responded to a survey of agricultural specialists.[15] Policymakers tended to see grain companies or supermarket chains as wanting the same things. For example, numerous individuals incorrectly identified Bunge and Continental with ConAgra's grain export subsidy plans only because they had seen these firms previously supportive of one another on other issues.

Trade associations are also susceptible to disagreements among member firms and, in several cases, member co-ops. Respondents from each trade organization surveyed for this study reported this problem and most of them acknowledged that avoiding controversial issues was the only way to avoid breakdowns within the association. So some of the issues requiring a firm's anonymity are also shunned by the trade associations. Not all potentially important issues can be addressed, unless members do support a consensus or a strong working majority emerges. The attitudes of CEOs about the appropriateness of addressing certain policies are limiting factors, too, especially since many executives shy away from agricultural production issues and urge the association to do so. The large-membership Grocery Manufacturers of America (GMA) demonstrated two distinct lobbying tactics on major issues of 1985 for that particular reason. With a united membership, the GMA was an aggressive leader in opposing a value-added tax for the Superfund. But, with a divided membership on farm support programs, the GMA issued only a mild public endorsement of market-oriented agricultural policy. The Corn Refiners Association (CRA) followed the same approach even though its membership of only eight firms might have made compromise easier on single-issue topics involving refining. The CRA's rule of thumb was easy: avoid issues like ethanol subsidies in which not all members are involved, always address issues such as commodity loan rates in which members' finances are affected and by which allied interests will not be too offended.

These same problems confront the producer-supported marketing organizations that also encourage agribusiness membership. Agribusinesses with trade interests find participation in these organizations advantageous in that their differences with producers can be addressed in a common forum. By carefully cultivating participants, demonstrating leadership, and negotiating compromises, agribusiness representatives can become integral parts of producer organizations. The most divisive issues are seldom solved, however. Agribusinesses must be content with such small gains as the structuring of international marketing offices and the use of these organizations as a neutral meeting ground. A few agribusiness representatives, including those from otherwise active companies, believe that their most important contributions to the firm result from this interorganizational lobbying with producer representatives.

UNOBTRUSIVE TACTICS

The biggest disadvantage of association-based lobbying comes from the lack of opportunity to create an immediate policy impact. There are few opportunities when conversions are rapid or when public relations campaigns are dramatic. Agribusiness lobbyists whose interests are not compatible with producers particularly lament the widespread public recognition of a farm crisis. With sympathies and electoral fears directed toward financially troubled farmers, trade issues and market problems have been difficult to promote successfully if they appear inconsistent with enhancing farm incomes immediately. To gain access for simply addressing such issues as lower loan rates, even in a climate of Reagan administration opposition to farm programs, several agribusiness representatives claim to have found it necessary to elevate their own status. Agricultural experts have been activated to raise the issues, and they frequently note that the farm crisis and commodity programs can be treated separately. Prominent business leaders, not necessarily from their own firms, have been encouraged to contact policymakers to make an electoral connection. Political Action Committees have increased in number and appeal because funding incumbents increases a firm or association's recognition.

Even with these tactics, however, agribusiness's collective policy approach has left these firms in a weak position as an integrating force in structuring agricultural policy. On production issues, whenever popular support is directed toward farmers, agribusiness occupies a dependent position vis-à-vis both policymakers and those organized interests to whom policymakers usually listen on farm matters. The legitimacy that policymakers accord farm groups on these issues, farm group complaints about the evils of agribusiness, the reluctance of many CEOs and agribusiness representatives to encroach on production aspects of policy, the divisions within agribusiness over appropriate policy directions for agriculture, and the willingness of most firms to make adjustments first in the marketplace and only then move to politics put agribusiness at a political disadvantage. The established farm lobby of the 1980s, on the contrary, benefits not only from its own aggressive strategy and immediate demands but also from the public attention and presence farmers gain from the turmoil of protest.

An indicator of this comparative disadvantage can be seen in legislative staff opinions of agricultural interests. Staff members were asked whether they recognized a sample list of 100 generally active organizations. An average of 38 organizations per respondent were not recognized.[16] While Washington-based farm organizations were always recognized, agribusiness firms and trade associations were recognized only 54 percent of the time. The importance of the data can be better determined by relating recognition to the respondents' perceptions of influence. Respondents were also asked to rank recognizable groups according to "How Active and Involved" and "How Influential" they perceived each organization to be on agricultural policy. Possible responses included "very," "somewhat," "a little," and "not at all." For Washington-based farm and agribusiness organizations, respondents recorded a 72 percent direct correspondence between levels of activity/involvement and influence. Almost all (19 percent) of the variation in the two responses for these interests involved a single category shift in response from one question to the next. When there was a variation in the response, influence was usually (92 percent) ranked lower than active/involvement.

This response pattern did not hold for agricultural protest groups and public interest lobbies, however. The protest organizations, when recognizable, were always ranked high in activity/involvement. Eighty-eight percent of the time, influence ranking declined by two or more categories. Public interest groups ranked lower in activity/involvement than did protest groups, but their influence consistently fell, too.

These findings suggest the importance of high visibility for organized interests and a good working relationship between those interests and policymakers. Established farm organizations were twice as likely to be ranked as very or somewhat active/involved as were recognized agribusinesses. Because active agribusiness interests were always rated as influential, though, it appears that lower-influence rankings are less meaningful than the perception that agribusinesses generally care less about being a prominent part of the process than do their farm counterparts. Some agribusinesses appear, on the basis of these data, to be highly esteemed as policymaking partners. The more muted agricultural policy involvement of agribusiness as a whole, however, makes it likely that many of these interests are dependent upon the policy fallout of relationships between other organizations and policymakers. In this case, caution hurts, because others do see political reticence as somewhat unseemly in an assertive policy environment. Harm comes mostly on mainstream issues of production agriculture and only as a result of a willingness of agribusiness leaders to adjust to its consequences rather than to fight.

CONFLICTING INTERESTS

Before making any judgments about agribusiness's role in agricultural policy-making and assessing the relationship between these interests and changing agricultural conditions, more careful attention should be paid to differences between individual firms and between agribusiness firms and producer interests. Much has been made of the belief that, despite differences within agribusiness, there remains a preferred businesslike way of structuring agricultural policy. The dynamics of both organizational and policy change fail to support that contention.

These differences in policy views within agribusiness were referred to frequently by farm and industry lobbyists and by the consultants they employ. Nearly the same fragmentation of interest in commodity policy exists for agribusiness as for farm organizations. Dairy, sugar, and peanut opposition by candy firms has been noted. Pizza Hut, a partner to candymakers in opposing dairy programs, has no interest or enthusiasm for challenging sugar programs and needlessly angering supporters. Antagonism toward producer groups on commodity policies is no more commonplace for agribusiness than is cooperation. Sometimes the supportive alliances are established firmly over time, as between the representatives of the Wheat Growers and the American Bakers Association, which agree on a "well understood, shared sense of common purpose." The Corn Refiners, its independently activist member, ADM, and the Corn Growers have an evolving relationship of that type. In neither alliance are commodity programs targeted for elimination. Users, however, have articulated the need for lower loan rates, but not at the expense of farmer income. No such sense of cooperation, common purpose, or concern for the other's economic well-being characterizes the relationships between candy manufacturers and dairy or sugar producers.

These different strategies evolve from both variations in the participants' viewpoints and the conditions that affect the industry. High-cost programs do not necessarily create enemies, but low-cost programs may well have many foes if the officials in user firms become opposed philosophically to interventionist farm economic policy. ADM President Dwayne Andreas, with his Minnesota Democratic-Farmer-Labor Party ties and background in the Farmers Union

Grain Terminal Association, behaves differently than many CEOs, at least in part because he learned the manipulative advantages of utilizing price support programs in influencing both markets and politics. The prospect for an all-encompassing agribusiness bandwagon on behalf of any single policy alternative is unlikely. Too many firms would see the threatened demise of those conventional farm programs that their executives have successfully used for price advantages, market expansion, and stable product supplies.

As might be expected, considering the long-term policy concerns of agribusiness and their executives and the individual predispositions to particular policies, the politics of agribusiness is not as dynamic as that of producers. While firms and trade associations come in and out of the policy process, they do so less regularly than do groups with more volatile memberships lacking in the institutional memories associated with businesses. Policymakers claim to know what proposals or actions will elicit responses from specific firms. Among grain traders, for instance, ConAgra can be counted on for advancing trade programs that require government support whenever foreign sales seriously lag. Cargill, in response, will disagree with ConAgra. Bunge Corporation, a firm whose representatives would go to Congress only under subpoena in the mid-1970s, is typically noncommittal over such disagreements, but if grain quality standards are brought up, it will quickly send out its attorneys. Producer groups, both lobbyists and policymakers believe, respond erratically, depending on member conflicts within the organization. Since trade associations tend to avoid issues of conflict and because so many individual firms want to maintain low political profiles, factional issues tend not to get played out by agribusiness, as farm groups are wont to do. "The farm guys have to take risks [in order] to be politically responsive. We do not [have to] and will not do so," said one company lobbyist.

Evidence of this attitude can be seen in the small number of instances in which agribusiness firms or associations publicly moved from one policy position into very different proactive representation. Respondents from all sources noted only fourteen examples of sharp departures in agribusiness policy involvement during the years 1980–1985. This contrasts with more than two hundred reversals of policy views from farm groups over the same period.[17] In each case, for agribusiness, these departures resulted from severe changes in business conditions or changing product involvement. Six firms and associations were newly activated over financial problems and responded in ways not consistent with their organizations's immediate historical involvement. The Fertilizer Institute (FI) is the best example, since it suddenly went all out in an expensive campaign against farm programs and tried to assume a broader political leadership position for agribusiness. The Irrigation Association, threatened by the same collapse in its industry's farm purchases as FI, dismissed nearly its entire staff and set out to develop a strategy for gaining political recognition of neglected industry problems previously addressed only by environmentalists.

Changes in product circumstances prompted eight similar reactions. Monsanto's involvement in farm issues, along with its financial support for the emerging Agricultural Research Institute, resulted from the company's expanding investments in biogenetic research with its crop production implications. ADM's move into ethanol production from corn prompted, or perhaps was prompted by, lobbying to expand government supports for the industry. Over the years, ADM obtained the biggest share of the federal government's billion-dollar direct subsidies and tax advantages for ethanol producers. Electronic Data Systems's (EDS) sudden interest in food stamp policy came from its ability and desire to develop an electronic distribution monitoring system for the program.

The most revealing policy shift was that of the Food Marketing Institute's slow and cautious movement toward eventual opposition to the dairy program. FMI members, through Department of Agriculture contacts, initially were asked to endorse and house federal dairy product distribution programs for the poor. While refusing to house the programs, FMI representatives saw little choice but to lend their goodwill in a recession environment. After carefully monitoring the program and supermarket sales of cheese and butter for several years, FMI eventually concluded that products were no longer going to just the needy. However, grocery market images were not thought to be benefited by a frontal attack on programs for the poor and the elderly. Instead, FMI elected to blame the dairy program for the massive dairy surpluses that had triggered the giveaways.

In addition to the generally reactive nature of these reversals in policy position, the examples all involved new sets of organizational actors as well. To an important extent, this demonstrates the significance of public policy entrepreneurs in business. In much the same way that group and association leaders have been found instrumental in creating political organizations, businesses appear to need their own entrepreneurs who create an internal momentum for changing the status quo of ongoing political transactions between firms and government officials who come to look for predictable company reactions.[18] Both the Fertilizer Institute and Irrigation Association changes were set in motion by member CEOs who were dissatisfied with the lack of organizational responses to farm programs and environmentalists' control of the public-issue agenda. Both associations hired directors and staff on the basis of the ideas these professionals proposed for corrective action. Operational plans were then left to the new employees. In the cases of Monsanto, ADM, and EDS, new product lines could develop only if supportive public policy conditions existed. An awareness of these necessary circumstances came only as the managers responsible for the products studied the total business environment of consumers and government. Food Marketing Institute officials arrived at their conclusions only through the input of new employees with USDA and legislative staff backgrounds who had a familiarity with farm programs that others lacked.

The internal sorting process, rather than the type of response decided upon, reflects the conservative character of agribusiness to which policymakers and lobbyists frequently allude. Neither new product lines nor new marketing strategies, especially if they involve a public policy component, win easily; complex organizations decide their futures. Participants in the process contribute winning suggestions when they are able to show a strong potential for profit to a firm's carefully maintained image without attendant losses. Losing proposals, usually subjected to the scrutiny of many executives from all major divisions within the company, generate no convincing proof of gain. If respondents can be believed, agribusiness decisions usually move in the incremental directions of reaffirming past practices. So, Monsanto does not propose an alternative farm bill as part of its mid-1980s strategy. John Deere does not view its lobbyist staff as a potential vehicle for reversing financial losses. Ralston-Purina rejects almost all proposals that bring the firm any prominence on specific rather than general policy positions unless they involve something narrow such as pet food regulation. Lobbyists perceive that these divisions result from the conservative practices that CEOs learn as they work their way through company ranks and eventually replace executives whom they then emulate. "Things would be different," said one long-time employee, "only if we hired an Iacocca or if there were more risk takers like Dwayne Andreas."

If the views of those who have worked for agribusiness are accurate, and their unanimity of opinion suggests that they are, there are significant implications for agricultural policy. Those who expect agribusiness to become an integrating force, or a broker, of agricultural policy are likely to be disappointed. Agribusiness interests are too diverse; and the conservative behavior that characterizes these firms, to the extent that it does exist, is one of caution more than of political philosophy. Executives and their political representatives remain pragmatists, not adverse to utilizing government support programs that benefit the firm or industry. In this sense, they reflect the conventional values of political activists, even though they tend to distance themselves more from the policy process.

The politics of farm bills appears an especially unlikely process for considerable agribusiness leadership. Risks are many and gains are few. Furthermore, farm bill politics exists in an especially rancorous and open environment in which major players go out of their way to attack opponents.[19] Robert H. Salisbury, in his analysis of the state of interest group theory, regards "conflict as problematic."[20] From agribusiness's perspective, this condition results from a specific desire to avoid controversy even though the benefits of a preferred outcome might be highly desirable if these rewards could be obtained in that totally concealed backroom of political fable. Taking risks appears too uncertain when company attention can be better directed to such heartland issues of industry as taxation, critical product issues before the Federal Trade Commission or Food and Drug Administration, or just manipulative marketing practices aimed at winning consumers.

To this end, agribusinesses can be counted on only to fill a number of primarily food and trade niches within the total array of agricultural policy problems. Firms shop selectively, often in conjunction with compatible producer interests, from among the array of advantageous provisions that might be worked into farm legislation. To an important extent, this attention provides coverage to issues such as commodity-clear title, which farm interests might not otherwise cover adequately. In other instances, however, agribusiness participation simply complicates the political process by adding side-payments to issues that might be better decided on the basis of the commodity problems of high budget costs and surplus products confronting policymakers. Corn-refining industry support of expensive sugar programs to benefit corn sweeteners is one costly example in which the winners actually gain by promoting market failure.

7

The Other Lobby:
Private Views
with a Public Interest

Should the many other interests that watch and react to agricultural policy adopt a collective slogan, it would almost certainly read: "What havoc has been wrought!" It would symbolize the interests' great difficulty in holding the attention of agricultural policymakers. This chapter examines a large number of organized interests that define their policy missions as distinctly different from those of farm and agribusiness lobbies. The conflicts that result from the incursion of these other organizations into agricultural policy is made more pronounced by differences in political style, lobbying strategies, and relationships with policymakers. These distinctions, in large part, are created by organizational matters. Farm and agribusiness interests are represented most actively by organizations that attend to a variety of policy problems, even though they may often have a single-issue focus. On the other hand, most of the active agricultural lobbying for these other interests is done by small, narrow, single-project organizations or similar units within more diffusely focused groups that are involved infrequently. The contrasts, as they are played out in vying for influence on what has long been considered farmers' policy turf, compound the irritations that these diverse private interest representatives feel.

What are the concerns of these other groups, some of which were referred to in Chapter 2 as public interests? Usually, in a pejorative way, policymakers with farm program responsibilities call them nonagriculturalists. That label fails to describe adequately Iowa consultant Roger Blobaum, who, in addition to his NFU and NFO background, has provided much of the policy direction for the National Catholic Rural Life Conference. Nor does it any longer fit Garth Youngberg, a former USDA official who heads the Institute for Alternative Agriculture, or Jim

Hightower, the controversial Texas agriculture commissioner, whose Agriculture Accountability Project of the 1970s did much to break the long-standing inertia forestalling critical attention to established agricultural practices and interests. But being labeled a nonagriculturalist is more a description of loyalties than of expertise. It means being consigned to that place of dishonor reserved for those who believe not in serving agriculture first but, rather, in forcing policy advocates to prove agriculture's contribution to society and its environment. Among traditional agricultural representatives, that contribution seems beyond dispute. Group representatives such as Blobaum, Youngberg, and Hightower, who have spent years involved in studying farm problems, believe on the contrary, that most assumptions about agriculture deserve critical review.

These interests are often referred to as citizen groups.[1] Don F. Hadwiger, a sympathetic observer of these interests, coined a more useful phrase than either *citizen group* or the title of *public interest group*. Because of their focus on policy, Hadwiger calls them *externalities/alternatives organizations*.[2] This descriptive title broadly summarizes their purpose as a set of reform-minded organizations that, depending on the group, represents either a form of "good-government" agriculture or the special desires of those who want more radical transformations in agricultural practices than do farmer and business interests.[3]

The ex/al groups are directed, first, toward the unintended side effects of a wide variety of agricultural policies. A policy result, for example, may provide producers a bigger market, input industries more profit, and growers and manufacturers more stable community supplies, but it can still produce undesirable results for those who eat the product or live near the producers. Consumer interests may well be displeased with Hightower's machine-picked Hard Tomato and the deterioration in food quality that it represents.[4] Environmental interests may oppose the concentration of chemicals left by intensive production methods. Conservationists may find the associated demise of terraces objectionable even if runoff and erosion damage no one's property but the farmer's own.

Both in public programs and agricultural activity, alternatives have emerged as essential policy goals for all of these interests, even if reform options are not specifically delineated. If tomatoes are tasteless or health hazards too great, reputable public policy demands can hardly mandate either no tomatoes or no modern farm practices as solutions. In this sense, the search for viable alternatives—at least as generalized suggestions—must be a consuming organizational pursuit, as these groups demonstrate that farmers and middlemen have not always been especially good neighbors. Many ex/al groups, as their representatives note, are more concerned with alternatives than with externalities, since these interests constantly confront the central but rhetorical defense of agriculturalists: How could inexpensive and plentiful food have been better supplied? Though some animal rights activists, for instance, see externalities affecting the human spirit in transactions that lead to animal slaughter, most simply want more humane techniques. Rural groups want better facilities or safeguards for farm workers only because they enhance both worker convenience and public health. These conditions, according to ex/al representatives, can no longer be tolerated just because they make food production efficient.

The focus of ex/al organizations on what comes about for others as both products and results of agricultural policy and practices, plus the emphasis on alternatives that may well force change, initially put these groups squarely at odds with farm, food, and fiber interests. While farm group complainants have often been critical of agribusiness, the representatives and financial supporters of both sets of interests are far more likely to reserve their real hostility for this Other Lobby of interlopers. Only over the past decade has any conciliation taken place, and then only because many of the issues addressed by ex/al groups have otherwise gained broad legitimacy in national politics and, through their support, proven useful in attaining the primary goals of established agricultural interests. Few of the interests yet find this newly shared policy ground an agricultural commons, though.

THE EMERGENCE OF AN EX/AL LOBBY

Andrew S. McFarland's conclusion that public-interest organizations are both old and new is an apt starting point for understanding the ex/al lobby within the context of agricultural policymaking.[5] Several of the most respected, best-supported, and—from a farm and food industry point of view—most feared ex/al groups are among the oldest and politically respected lobbies in Washington. The American Forestry Association (AFA) was founded in 1875, the Boone and Crockett Club in 1887, the Sierra Club in 1892, the National Audubon Society in 1905, the North American Wildlife Foundation in 1911, the Nature Conservancy in 1917, the Izaak Walton League of America in 1922, the Wilderness Society in 1935, and the National Wildlife Federation in 1936.

For each of these groups wilderness protection is the only common denominator, and land use and resource ethics remain prime agricultural concerns. Since its inception the Wilderness Society has addressed the question of the use of public land, especially the control of open livestock grazing. The AFA's opposition to the destruction of forests and the resulting soil and water resource losses are among the problems the group constantly monitors. As a consequence, farming practices periodically receive legislative heat from AFA policy demands. The historical position of each of these organizations evolved from the assumption that farm practices, in the main, blatantly disregard any natural resource use other than extraction. The basis for a collective understanding between these large but otherwise diverse groups can be seen in what their representatives have charged as frivolous inattention to the destruction of songbirds by chemical spraying, game through habitat conversion, gamefish through erosion and water pollution, recreational and hunting rights through producer encroachment, and forests through unnecessary clearance of environmentally fragile land. At the height of discussions between conservation and farm organizations over land diversion provisions of the 1985 farm bill, an ex/al lobbyist with a farm background who is now employed by one of these organizations voiced the lingering distrust of producer groups by exclaiming, "Carp! Farmers think a river is clean if carp can live there. Farmers can be counted on to fight like hell for you when it finally gets to the point where all we have left to protect is carp!"

While that assertion seems far from accurate, perceptions are often everything in lobbying when group representatives appeal, often emotionally, to a policymaker's best judgment. Because a tradition of unshared interests separates mainstream agricultural organizations from those ex/al groups that first paid attention to the ramifications of farm practices, cooperation and mutual trust have been difficult to foster. In a modern farm crisis in which producer groups see the need for some form of greater farm income enhancement, wilderness ex/al organizations see prospects for public policies that encourage land retirement, habitat restoration, and expanded recreational opportunities. That is, they see benefits in restricted farming. Moreover, because of what wilderness representatives and many members see as producer group motives that promote farming at any cost, the wilderness lobby has long sought to rally the broad-based support of other environmentalists, recreational industries, professional societies, state tourism interests, and whatever good-government forces they can recruit before venturing to participate in discussions involving farm lobbyists.

Wilderness groups, initially elitist, and their conservation positions have fared little better when evaluated by agricultural interests. Neither rural residents nor blue-collar workers numbered many members of such groups among their ranks, although the National Wildlife Federation earned most of its early funding through one-dollar contributions sent as compensation for its mass-mailed conservation stamps. Supporters have long been portrayed by opponents as city residents with the financial luxury of being able to trek to the countryside and expect an unused environment. The implication, often stated directly, was that hunters, fishers, and outdoor enthusiasts had no practical sense of economic reality in the face of an accelerating demand to feed a hungry world. In the 1980s, much of the expansion of ex/al groups is attributed by their staffs to the generosity of that percentage of the public having excess income and a willingness to spend it on aesthetics. Such supporters represent elitist values in the same sense that, as individuals, they are much the same population targeted by the food industry as non-price-conscious consumers. With both conservationists and food industries selling chemical safety and wholesome life-styles, a complementary policy approach provides some bonds between them.

Opponents cannot argue that ex/al interests and the values they represent lack political clout, however. The degree to which other interests rally around them is only one indicator. Each of these multipurpose organizations is involved extensively in an array of policy, public relations, and education matters that gain them widespread recognition. Perceptions of success are so high, in fact, that the National Wildlife Federation was ranked by independent judges as one of the two most powerful interests in Washington in 1984.[6] All of these organizations can count on well-placed policy spokespersons within government.

Thus, while an important part of the ex/al lobby is old, it represents a contemporary sense of values prized increasingly in a population that has more leisure time and, in a great many instances, more expendable income both to enjoy and use in protecting noneconomic values. Having pioneered this venture in lobbying for the noneconomic interests of an only somewhat assignable set of supporters, the conservationists and the fur, fish, and game interests that made up the wilderness lobby became a model, at least in terms of policy mission and attitude, for other ex/al groups that later addressed agriculture. Important additions, both in types of issues and social problems of concern, were made to this Other Lobby continuously after World War II, with a sudden escalation beginning in the late 1960s.[7]

While considerable research shows that entrepreneurial personalities have played an important role in founding and sustaining ex/al groups that become active on agriculturally relevant issues, the type of issue involvement varies; not all ex/al issues are being addressed at once. It hardly seems accidental that wilderness groups first emerged as frontier-style opportunities decreased for an urbanizing population. With westward opportunities too vast to contemplate, neither Daniel Boone nor Davy Crockett would have been much interested in the Boone and Crockett Club's record keeping or in its sponsorship of wildlife research. There was always something bigger in the next valley, it seemed. At some point, however, there were no new valleys, and it became time to protect those that wilderness proponents valued most. But proponents could not address all preservationist issues at once. More than other interests, ex/al groups have always had to wait until the time was right for new policy windows to open or for others to appreciate what their representatives and supporters understood and valued.[8]

Policy windows, or the opportunities for advocates of dissenting policy positions to promote their reform proposals, open only when conditions make them politically acceptable. Though interest groups may foster and even create these conditions, ex/al lobbyists contend that they must spend most of their time simply waiting for such windows to open when some event underscores the importance of their group. The late Clay Cochran of Rural America and other antipoverty organizations frequently described the plight of ex/al organization to his peers. He was fond of likening the policy process to a growing bed of coral. Neither coral nor public policy is seeded to produce a desired product upon demand. Instead, different environmental conditions give rise to the development of different types of coral or to different kinds of policy outcomes. For organizations dependent on public sympathy for a base of support, the coral analogy appears especially accurate as both a description of how policy windows open and what ex/al activists must do while they wait. As one ex/al organizer and strategist wrote of the hard work that necessarily goes into coping with these fleeting opportunities to promote change: "It means not accepting simplistic either/or formulations of complicated human situations . . . it means being concerned with effective activity rather than purity of the line."[9] He concluded by observing that the reformist who does not wish to accept the externalities of others' decisions must, "change his identity to fit all new situations."[10] In short, to mount an impressive lobbying effort on a winnable issue, ex/al representatives find it more effective to address and expand already emerging issues rather than work on only personal priorities that neither the public nor policymakers may comprehend.

The remarks about effective action emphasize how important the opening of policy windows has been to an expansion of the ex/al lobby from its initial handful of wilderness groups to a proliferation of mostly small single-project interests. The entrepreneurs who created these organizations, frequently on small grants, moved into the policy arena on the basis of issues and ideas that were marketable to funding sources.[11] Through 1976, most of these groups operated under the Internal Revenue Code as 501(c)(3) organizations with a restricted ability to lobby. Under the Tax Reform Act of 1976, permissible lobbying activity was expanded even if funding sources were from private foundations.[12] Prior to tax reform, ex/al groups that qualified for 501(c)(3) status lobbied gingerly, with an emphasis on allowable advocacy within public agencies and the distribution of analytical research. A few organizations operated separate corporations to receive both foundation support and government funding legally, while the parent group lobbied more actively.

The problem for ex/al organizations operating under the old code was obvious. Effective lobbies, those that could act like farm and agribusiness interests, either had to have an extensive and supportive general membership that allowed them to forgo 501(c)(3) status, or they had to operate within strict limits. Under 501(c)(3) restrictions, only a quite large and well-funded ex/al organization could muster the resources to gain an extensive degree of political access. Newly emerging interests rarely become large. The organizations learned to operate from small offices, with small staffs, and by selecting carefully what issues to target. For those who governed farm, food, and foreign trade policies of agriculture, the ex/al lobby of new interests developed as a bothersome but not generally ongoing presence. As long as wilderness issues could be negotiated or avoided, agricultural policymakers faced little more than a trickle of single-project organizations whose representatives had little experience in agriculture and who only occasionally used open policy windows when issues of concern to farm or agribusiness groups were already in the public eye.

Through the 1950s and well into the 1960s, the expansion of ex/al groups continued to be primarily along traditional conservation lines. Many were regional organizations with protectionist goals for unique land areas or, in a few instances, animal species. Their strategies varied according to local problems and the unique preferences of group activists. ACRES in Indiana, for example, set up public land trusts, and the Big Thicket Association of Texas initially directed its efforts to education and information. The Natural Areas Council emerged in 1957 to provide direction and assistance to regional efforts; but, like its local counterparts, its alternative goals of placing private lands in trust needed little assistance through public policy reforms.

Politics and public policy were, however, the concerns of a variety of groups that began to become active in agricultural issues in the late 1960s and early 1970s. Their diversified interests meant that established agricultural groups encountered these new opponents more frequently on more and more issues. Rural interests, most notably Rural America, organized out of the civil rights and anti-poverty movements of the 1960s as activists became convinced that USDA was more attentive to its farm missions than to its rural responsibilities. As attention to rural issues expanded, groups such as the Rural Coalition, Rural Governments Coalition, and National Rural Housing Coalition organized to work for increased federal aid for nonfarm rural constituents.

Consumer discontent, hunger problems, and world food supplies served as organizing issues for several groups. The environmental movement, as a sharp departure from the recreational concerns of most wilderness groups, became a new conservation force protesting the effects of human use of natural resources.[13] Later in the 1970s, animal welfare activists turned their attention to farm animals. Several animal rights groups were well established and to some extent already in conflict with producers over the killing of livestock predators. While five to ten public interest organizations were formed in each of these five issue areas, only the Agribusiness Accountability Project (AAP) was organized specifically around agricultural production. The AAP, however, had little in common with the small number of rural interests that had emerged with a nonfarm policy focus.

Ex/al representatives see several reasons why these interests eventually moved to address agricultural policy. First, the public interest movement in cast and character evolved from the activism of civil rights, antipoverty, and peace causes.[14] Like many agrarian protesters of the 1980s, the young organizers shared neopopulist beliefs about the need for countervailing political power based on citizen input. Second, either programs were in place, or changing producer conditions had already focused attention on the implications of agricultural issues consistent with neopopulist causes. Domestic and international food assistance programs were operating, but they had been designed as programs that helped eliminate production surpluses.[15] The rural hunger issue, as articulated nationally through Robert Kennedy's Senate poverty subcommittee hearings, directed both attention and conflict to the lack of client concern in food stamp, commodity distribution, and school lunch programs. World hunger led many to question the adequacy and the intentions of U.S. food assistance, distribution, and school lunch programs. For a time, health concerns joined consumer and nutrition interests.

Sharp price increases and greater middle-class frustration later led to a distinct consumer interest, separate from the hunger lobby, as cooperative members of the Consumer's Federation of America pressured the organization to redirect its attention to marketing issues. For environmental and animal rights activists, structural changes in production with increased reliance on chemical inputs provided a natural bridge from general issues of toxicity and cruelty to specific methods of growing crops and raising livestock in confinement and with extensive use of preventive antibiotics. The Center for Science in the Public Interest, with its research emphasis on the effects of pollution, was instrumental in pointing to linkages between environmental pollution and health hazards. The major work of the Agribusiness Accountability Project criticized land grant universities as primarily responsive to the same corporate food industry that cared little for nutritional value, consumer needs, production methods, or family farmers.[16] Rural interests, through the leadership of antipoverty activists like Clay Cochran, eventually argued for development and assistance to rural communities as a means of providing world food needs, lowering food costs, serving family farmers, and providing the jobs needed to eliminate poverty and rural hunger.

The surrounding conflict and political attention created by these issues not only opened policy windows of great social importance to activists who were looking for policy roles; it also opened windows where small but convergent interests could temporarily put aside differences and work together to share scarce resources on whatever issue was before the public. If placed together in the context of corporate and large scale producer irresponsibility, the issues that these groups addressed were closely related. Assuming irresponsibility and further relating that assumption to the neopopulist need to redirect political power to a people's movement meant that the new ex/al groups entered the political process even more antagonistically than wilderness groups had done. Public interest rhetoric could hardly have been both softly articulated and still credible during this period of development, and zealous organizers probably would not have wanted to make cooperative overtures to farm and agribusiness interest anyway. Litigation was not only effective in the civil rights movement, but it was also appealing because of its adversarial overtones and winner-take-all judgments. Established agricultural interests and policymakers became overtly hostile to these new participants who suddenly appeared in the midst of already existing debates to carry issues further than anyone else. These new activists were disliked because they played upon often minor internal disagreements between established policy activists and then, through rallying public sentiment, imposed on such institutions as the Department of Agriculture and Food and Drug Administration new agenda responsibilities that would otherwise have been avoided.

Public interest groups that newly organized to tackle agriculture during the

1970s, according to participants, did not get established without the support of other ex/al interests. Religious and labor groups were instrumental in providing funding and strategic advice, and they continue to be important parts of this Other Lobby.[17] These are essentially liberal organizations that earlier had supported civil rights and antipoverty causes. Their members, whom agriculturalists saw as "the original social do-gooders," had even more traditionally defined self-interests in safe, high-quality foods being made available to the nation and to the world's needy. As a spin-off of this support, in-house agricultural units were added to several religious organizations that had lacked them. After years of involvement in world food issues one such organization, Interfaith Action for Economic Justice, became a foremost advocate of targeting commodity price support programs to only those farmers who had a demonstrated financial need. Rural and labor interests also lent considerable assistance to those groups that supported migrant workers, especially the United Farm Workers with their prominent base of Washington supporters.

Although specific organizations would form and disappear, the essential structure of the ex/al lobby was set by 1975. Issues have been relatively constant since a new policy agenda was first recognized by USDA officials, who were more attuned to production and trade policies. From a lobbying perspective, that new agenda now means that interest representatives can be expected to make demands at any point where new policy windows open opportunities for articulating issues of land use, acceptable agricultural inputs, product practices, and food manufacturing as these issues involve conservation, preservation, animal protection, consumer prices, food safety, nutrition and health, social conditions, and rural development. For example, animal rights activists entered the controversy between cattle and dairy interests over the implementation of the whole-herd buyout program. By challenging facial branding, that program's elevated visibility gave the Humane Society of America an excellent forum for raising its broader concerns about farm animal treatment. This kind of incursion by these constantly evolving interests has permanently affected both the influence and representation of interests within agricultural policy.

CHANGES WITHIN THE EX/AL LOBBY

Some things have remained constant for these groups. Their goals are still of the same ex/al type, the issues that involve them politically are similar, and an adversarial orientation toward most other agricultural interests prevails. No other ex/al interests have developed the multipurpose organizational capacity, at least in the units that address agricultural policy, of many of the older wilderness groups. The flexibility and lack of organizational restrictiveness found in small operational structures is a characteristic especially prized by most ex/al lobbyists. Staff members of the public interest groups that formed in the 1960s and

1970s retain their zealous conviction. In many instances, however, years of bargaining experience have softened their attitudes if not their suspicions toward other groups. Conservation representatives, in particular, have sought to minimize hostility in their relationships with other agricultural lobbies. Newer public interest groups, on the other hand, tend to be every bit as contentious as their predecessors, sometimes to the point of purposefully isolating themselves from interaction with those who fail to agree totally with their views. Many of these groups are political action groups that never come to Washington but promote local activism because of their negative views of national politics and their desire to be exclusively a community or education service.

As it became apparent that ex/al issues were permanent, still more ex/al groups were formed for specific conciliatory reasons with farm and agribusiness support. The express intent of these organizations is to address scientific and technical questions about the externalities of food production in the name of corporate responsibility towards customers. These groups are capable of engaging the community of long-established agricultural interests in an ongoing dialogue about policy alternatives. From the perspective of producer, agribusiness, and ex/al representatives, the Other Lobby has fragmented from its early image of cooperative social activists seeking redistributive political power. The splintering that once separated only the new participants from the old conservation groups now, depending on the perspective of various ex/al activists, includes antitechnological preservationists, agricultural adversaries, middle-of-the-road mediators, co-opted interests, and industry apologists. In short, despite its capacity and inclination for cooperative action, the ex/al lobby functions much like the rest of the agricultural lobby as a set of often vaguely related but heterogeneous interests. Due to changes in Internal Revenue Code 501(c)(3), members of the Other Lobby can now participate politically, and many have chosen to emulate other interests in style and tactics.[18]

Other notable changes have come about as these interests have matured collectively into an ongoing part of the agricultural policy process. Some groups have a distinct agricultural production interest. Many small groups have become involved. Groups specialize more than they did previously. Both industry and government are more comprehensively monitored, and policymakers expect and even cultivate ex/al participation.

OTHER ORGANIZATIONS

At the time agricultural policymakers were first commenting on their new policy agenda, the one organization that dealt with farm problems was primarily an anticorporate muckracker. The Agribusiness Accountability Project (AAP) was founded in 1970 by the year-old Center for Community Change and the only somewhat older Center for Corporate Responsibility. The AAP's family farm advocacy was prompted by its attacks on agribusiness corporations profiteering from land grant college research and Soviet wheat sales, both at the alleged expense of farmers. Its last hurrah, before funding died out and remnants of the group moved to San Francisco, was a joint Food Action Campaign for antitrust action against food manufacturers. Labor, church, and consumer interests supported a campaign that did little more than institutionalize the phrase "structure of agriculture" with those who would later occupy agricultural policy positions under the Carter administration.

Given the price instability problems plaguing agriculture in the middle to late 1970s, it seemed surprising that no organization quickly filled the void left by the AAP's demise. Farmers of all sort were left to speak for and organize themselves, especially through the American Agriculture Movement (AAM) and farm crisis committees, until large numbers of social activists began to be attracted to the agrarian reform efforts in 1985 and 1986. Given the reluctance of these newer ex/al groups to work for assignable policy benefits, this lack of earlier cooperation should not have been surprising. Given their problems with agriculturalists, they were no more interested in promoting farmer power with resulting producer rewards than they were willing to accept corporate power. As one ex/al lobbyist who still refuses to support farm protesters said, "Farmers do not represent the people of this country."

The rhetoric of the farm protest movement, especially with its anticorporate theme, eventually attracted the attention if not the active support of many of the social activists who saw something worthwhile to emulate in the conservation, consumer, and other public interest organizations involved with agricultural issues. AAM farmers, with neopopulist goals of their own, were fighting for redistributive justice in much the same way that blacks and students won substantive gains. Agricultural observers who had nothing to gain but the collectively assigned benefits of reform would study the production practices that producer activists would not address. AAM President Marvin Meek reaffirmed the need to address more directly agricultural issues on a broader basis when he acknowledged that he and others could see the negative environmental consequences of their production techniques but could not afford to deal with them because of farm financial conditions.

By the late 1970s and through the early 1980s, a farm-specific ex/al lobby was taking shape, mostly independent of the farm protest movement. Several organizations, some new and others old, began to investigate the impact of the results of large-scale farming. The Center for Rural Affairs became one of the few organizations identified with both ex/al goals and active support for agrarian reform policy. The center, previously a rural development organization, became an advocate for family farmers as critical links in small towns. Rodale Press began its Cornucopia Project to examine the prospects of sustaining high rates of production under current farm practices.[19] Rodale Research Center then helped establish the Regenerative Agriculture Association to investigate chemical-free, natural-energy-based, localized food production and distribution systems. The

Institute for Alternative Agriculture was set up just outside Washington, D.C., to merge organic and chemical means of food production under the scrutiny of agricultural scientists. The conservationist American Farmland Trust was formed in 1980 to protect farmland from erosion and development. Two other organizations were active under corporate, land grant college, and USDA sponsorship. The Council of Agricultural Science and Technology (CAST) was established to provide scientific inquiry into subjects of technical controversy associated with capital-intensive agriculture. CAST's isolation at Iowa State University, cumbersome organizational structure, and internal disagreements over its advocacy role led to the reorganization of the nearly dormant Agricultural Research Institute (ARI).[20]

Like CAST, ARI aims to mediate conflicts in areas of technological disagreement. Supporters, including agribusinesses, hope that ARI can serve as a voice for agricultural scientists who want to express their views from an organization within the Washington area and with a proproduction orientation.

The biological aspects of agriculture have received as much attention as have structural issues of farming. Biotechnology and genetic engineering have strong advocates within the scientific community, but the enthusiasm of these proponents has met with strong counteraction. Both the number of organizations attracted to policy questions of biotechnology and the range of concerns tied to their activism reveal much about contemporary ex/al specialization and the limits of intergroup support. Three separate organizational efforts have made substantial progress. The most comprehensive effort involved extensive research into the possible consequences of agricultural collapse in the face of the genetic failure of seed supplies.[21] Sponsored by the Environmental Policy Institute (EPI), the project's heavy demands kept EPI's agricultural staff so preoccupied that the organization, to the surprise of many, played no role in 1985 farm bill discussions. Environmental damage from experiments with genetically altered organisms prompted a series of legal suits by the Foundation on Economic Trends during this same period. Despite its expansive title, foundation representatives made these actions and community mobilization their sole organizational purpose. Other groups, organized on both a national and regional basis, responded to fears of environmental damage and food shortages by establishing centers for the preservation of unaltered seed supplies.

Despite the common concerns of these different organizations, lobbying efforts were not coordinated, nor did the groups interact routinely. Participants did little more than exchange information. This led to similar critiques of U.S. agriculture as organizers, proponents of land stewardship, antipollution activists, and family farm organizations shared ideas. Legal challenges linked biogenetic agriculture to large-scale production, inadequate attention to conservation practices, and a deepening financial crisis for small and medium-sized farms. Because of its opposition to farm animal confinement, the Humane Society of America joined the plaintiffs when animal experiments were the subject of foundation lawsuits.

GOVERNMENTAL RELATIONS

The reactions to biogenetic research illustrate the style of lobbying that small ex/al organizations with intensively defined interests may elect to follow. Typically, ex/al representatives maintained that their jobs demanded that they provide hard questions rather than definitive answers, even in addressing policy alternatives. In a strict sense, these lobbyists saw themselves as obstructionists because of the importance they placed on minimizing the likelihood of disastrous consequences of such events as the collapse of new genetic seed strains or the spread of rural hunger. As advocates of regulation, these representatives were in the agricultural policy arena to demand caution and enforce rigorous standards. Such activities demanded little in the way of conventional lobbying resources, usually little or no direct political access, and no knowledge of agricultural production. Either the courts or public opinion could be used to tie up the resolution of issues if effective arguments could gain attention, especially through use of the media.

Under these circumstances, available organization resources can be directed toward a careful and thorough monitoring of legislative proposals, administrative rulings, and public and private research projects. Decisions to intervene can be made quite inexpensively whenever an organization decides that something can or should be countered or checked. Even when its warnings are highly speculative, an organization following this strategy has little to lose. Policymakers are confronted with situations in which the ex/al organizations can acknowledge limited risks but still exert tremendous influence. Not only are these externalities to be avoided on their own merits, but potential public awareness of even the most unlikely results may become a political issue if ex/al representatives can attract substantial media attention. "The idea," said one lobbyist, "is to make sure public officials cannot act until proponents prove their case. Public attention is our best ally in this regard."

Obstructionist policy tactics and accompanying adversarial relationships with more established agricultural interests do not necessarily translate into a lack of political legitimacy, however. Observers and many policy participants from within agriculture too frequently conclude that no one would willingly listen to or believe ex/al views. Those who look specifically at the speculative nature of many ex/al demands, the lack of their representatives' training in agricultural fields, and the seemingly minor complaints that many groups have with a major reputable industry underestimate the political appeal of the Other Lobby.

Such a viewpoint obscures the reasons why those who urge caution and restraint in selecting policy options have won, perhaps grudgingly, widespread acceptance within government. Agricultural policy, as it exists in a broader context, is no less subject to the suspicions that follow the rest of the government from a post-Watergate, post-Vietnam, procitizen, and proenvironmental era. The agrarian protest movement proved that point and capitalized on it. Because the larger issues of social accountability and fiscal responsibility are institutionalized in politics, especially, in the mind of the public, the ex/al lobby in agriculture, though often criticized by frustrated policymakers, is not ignored. Ex/al groups, alone within the agricultural lobby, raise those issues and, in so doing, raise their own political status, as well. At least, Washington respondents believe that to be the case.

These groups also make tangible contributions to the policy process that even many USDA officials now acknowledge.[22] Given their watchdog concerns and careful scrutiny of what others are doing in their areas of expertise, many ex/al interests have the only comprehensive knowledge of current activities and projects within both government agencies and private corporations. They also have reputations for pinpointing problems within congressional districts that might be embarrassing electoral hazards for incumbent legislators, such as chemical releases in a local river. Such organizations, particularly local and regional ones, can also prove useful by helping to identify gaps in public-policy benefits that may be interesting to an individual legislator in either district or committee work. Focus Hope, a Detroit group, was able to focus congressional attention on the nutritional needs of the elderly. The result was a pilot program to distribute surplus food to the aged. The program was later expanded nationally.

Since 1980, public programs have been increasingly scrutinized for fiscal cuts, and many ex/al groups representing consumer and environmental interests have been singled out for assistance in identifying and challenging high-cost, low-benefit, or flawed public programs. With the costs of agricultural programs escalating and with the unresolved producer and agribusiness complaints about the direction of agricultural policy remaining, the ex/al lobby has shown some potential for adding useful information to decision making. American Farmland Trust, with its plan for linking soil conservation and fiscal savings, played an important role in the 1985 farm bill in just this way. Advocating legislation to penalize producers who would continue to add more acres to crop production by sodbusting and swampbusting in environmentally fragile lands, Farmland Trust and its allies worked major provisions into that bill that policymakers valued for their cost-reducing impact. Consumer groups, especially Public Voice, were similarly useful in directing political attention to the high budget costs and large producer subsidies found in dairy and sugar programs. As one legislative staffer concluded, "These people talk about things that otherwise don't come through to me."

PRODUCER AND AGRIBUSINESS SUPPORT

Hostility toward the ex/al lobby is easily discovered among agriculturalists, even in Congress. Most of it is highly generalized and directed toward "those who don't know agriculture," a phrase repeated endlessly by farm protest, producer, agribusiness respondents, and public officials during project interviews. The phrase has become the litany of long-time agricultural policy participants besieged by unrecognizable faces who, in a time of farm crisis, want to influence a transitional agriculture. Hostility escalates because, while more traditional participants are divided on the severity of the crisis, they tend to focus collectively on incremental change as a solution and believe that either financial assistance or freer market policies can restore a healthy U.S. agriculture. Neopopulist farm protesters are not much different in their opinions. The rhetoric of the ex/al lobby, as diverse as it is, often disputes that logic.

Since it is both so harsh and so frequent, rhetorical generalization directed toward the agricultural establishment of farm organizations, corporations, scientists, and policymakers evokes the greatest anger. Those who speak on behalf of farm workers, family farmers, consumers, the environment, animals, and labor usually place these forces collectively at odds with that establishment in an ideological or even a class struggle.[23] Corporate supporters who want to control production, genetic seed supplies, energy, the land, or political power are often portrayed at the control center of an evil empire. Removing them and their biased agriculturalist supporters is a popular neopopulist mandate that is, despite other differences, shared with agrarian protesters. With only 1.4 percent of U.S. farmland held by nonfamily corporate ownership, agricultural policy participants whose own biases almost always favor family farming become enraged at the charges, those who make them, and those interests on behalf of whom the charges are leveled.[24] Consequently, there is little common ground. The central disagreements over the concentration of land and wealth have agriculturalists agreeing on the need to produce efficiently and the vocal ex/al groups challenging the by-products of economy and efficiency.

Unfortunately, the effects of this rhetoric upon those considered part of the establishment are often reinforced by specific experiences. When asked to comment on ex/al representatives, most other respondents cited obstructionist tactics employed by specific ex/al groups that refused to negotiate, discuss the issue at hand, or prove their own emotionally charged cases. The Industrial Biotechnology Association sees the Other Lobby as a supportive extension of the Foundation on Economic Trends. The National Cattlemen's Association looks at ex/al interests as a collection first of animal rights activists who oppose both feedlot containment and the control of predators and, second, wilderness representatives who want to eliminate cattle grazing on public lands. To the Fertilizer Institute, an ex/al lobbyist appears as an organic farmer with a judicial restraining order in hand. The repetitive cycle of adversarial confrontations between established and ex/al interests produces a situation where everyone remembers—and talks to friends about—unyielding conflicts that take on a larger-than-life symbolic importance. Under these circumstances, an agribusiness lobbyist can become livid at the mention of environmentalists generally, but later acknowledge with some pride his personal contributions to the Animal Protection Institute.

For most interests, the barrier to coexistence is more illusory than real. While the Humane Society of America's Michael Fox and the Foundation on Economic Trends' Jeremy Rifkin are unyielding in their opposition to poison baits and biogenetic experimentation, respectively, other opponents do work out many differences. For example, the heavily contested regulation of pesticides divided the entire agricultural lobby from 1984 to 1986. Agrichemical companies and their trade association representatives were finally able to compromise not only with environmental groups but also with representatives of labor and consumers.[25] An extensive series of working meetings produced agreements for chemical safety and funding costs for product reviews. The issue of cargo preference shipping tied up farm bill proceedings for several months in 1985 until labor and some agricultural organizations could negotiate the basis for a compromise. In preparing for the same bill, Public Voice worked with both the Farm Bureau and the Food Marketing Institute. The American Farmland Trust contributed key conservation elements of that bill with the support of several farm groups, especially the NFO and the NFU. It is, in fact, increasingly difficult to find examples of issue areas where the most visible ex/al groups have not reached some major issue compromises with established farm and agribusiness groups. Even animal rights activists, usually seen by respondents as the most militant and unyielding of those in the Other Lobby, ultimately agreed with land grant college representatives on inspection regulations for laboratory animals.

Many of these agreements, of course, are negotiated by the participants or outside brokers for no other reason than to produce a legislative or administrative outcome after a long stalemate. At those times, respondents acknowledge that all parties may feel like losers, even though partial gains have been made. At other times, despite the sentiments that continue to divide the two sides, established interests have become supportive of ex/al positions and gained from the articulation of their demands. Even conservative farm lobbyists consider conservation provisions, nutrition programs, and environmental regulations to be in the long-term interests of their members. After listening to the anticorporate criticisms of an ex/al advocate during a private foundation program, one farm lobbyist responded that, while he could not agree with the complaints, he did agree with the importance of articulating those concerns continually. "Somebody has to keep saying them," he concluded. The reasons for this support are many: ex/al input helps avoid critical errors, these groups consider variables that farm and agribusiness interests are not attuned to, and they provide long-range analysis of farm problems. "It gives us a warning," concluded another farm lobbyist. "That's why we talk to these people. We need to know what they know to protect our own credibility. We listen [to ex/al groups] more than people think. Doesn't everyone really want an abundant supply of safe food in the future? Of course."

While many respondents voiced support for establishing relationships between ex/al and other agricultural interests, the actual degree of interaction was not high. Most commodity groups and agribusinesses merely monitored these newer organizations to determine what kinds of policy problems they might produce. The general farm organizations, larger multipurpose commodity groups like the National Cattlemen, and the larger multipurpose trade association were the most willing to seek contact. Even then, direct contact with only what were often referred to as "more moderate" ex/al groups was felt to be worthwhile. Public Voice, Bread for the World, the Environmental Policy Institute, the Food Research and Action Center, and the American Farmland Trust were most frequently mentioned as having useful information and responsive representatives—at least for some established groups. Though this indicates slowly changing relationships, even this limited support represents some new directions in the representation of what matters most to U.S. farm and food interests.

CREATING A POLICY IMPACT

Even though the intervention of ex/al interests in agricultural issues has been far more limited than that of either producer or agribusiness lobbies, the collective importance of these groups in policy deliberations remains. Organizations such as Public Voice or the Environmental Policy Institute are little different from most other agricultural interest groups in facing problems of recognition, credibility, and political acceptance as they, like so many other interests, involve themselves in policy questions only periodically. The mutual support of well-funded and prominent sister organizations such as the National Wildlife Federation (NWF) offsets many of the disadvantages of limited political access. The NWF can use its publications to promote the viewpoints and findings of less resourceful groups and, in the process, cover an issue that its own staff has neither the time nor the expertise to research and address. As a result, the causes of many small ex/al groups can gain the same stature and legitimacy as issues promoted by the ever active American Farm Bureau Federation or the Food Marketing Institute.

The nature of the issues addressed by this Other Lobby ultimately does more to distinguish these organizations from either agrarian protest groups or established agricultural interests than does smaller size, more limited funding, or their greater tendency to exist as single-project organizations, even though these are all contributing factors to what issues they address and how they do it. Despite the wide range of issues that farm and agribusiness groups address, their interests are essentially economic. Questions of social justice, environmental need, and nutritional value may be given some perfunctory public attention, but the search for financial advantage structures the lobbying agenda of farm and agribusiness interests. When protest groups emerge around problems of market failure or when established interests address that breakdown, the search for corrective action centers around changes in taxes, trade, or price supports. Ex/al groups occupy positions of strategic importance in agricultural policy matters precisely because they are the only ones that look at market failures and proposals for corrective action from an entirely different perspective, one of social reform. In addressing policymakers, most agricultural interests have focused on economic choices about how to allocate resources and make efficient use of them. By so doing, producer and agribusiness groups have left others to ask the more basic questions as to which resources agriculture should use and under what conditions those resources should be allocated to food and fiber production and manufacture.

Under such conditions, when appropriate policy windows open, individual ex/al groups have amassed great political influence, albeit on a usually narrow issue. Sometimes, as has been the case with laboratory animals, organizations have had only to raise the question of improper care and direct public attention to cases of mistreatment to find themselves with an issue that no one can push off the policy agenda. Legal provisions and sometimes protective administrative institutions are put in place to help ensure that the initial issue is not overlooked, even though the group that raised it may have directed its attention elsewhere or may no longer exist.[26] While environmental organizations remain vigilant over agrichemical use and now negotiate long and hard on such issues as FIFRA, the compliance mechanisms that these groups helped to create earlier through the Environmental Protection Agency do more to ensure that questions of social choice will be addressed than do the continued research and public relations work of environmental activists.

The history of recent farm bills indicates that organizational sustainability and the continued articulation of policy demands are far less important than the initial impact of emergent ex/al participation. After 1973, following ten years of disagreement over the continued inclusion of the food stamp program in the ommibus bills, the program was so well institutionalized that no interest group initiative was necessary to protect its inclusion. Legislative tinkering focused on issues of general welfare reform such as workfare or matters of economic self-interest—at least to the snack food industry—such as junk food purchases.[27] The 1985 farm bill saw widespread support for conservation measures and the establishment of provisions that conservationists had worked for over many years. The cost-saving advantages of these provisions, along with the acknowledgment that fragile lands need preferential treatment, make it highly unlikely that sodbuster and swampbuster prohibitions can soon be reversed. American Farmland Trust and its allies, which for many years pronounced such provisions necessary for U.S. agriculture, can move on, at least to some extent, to other matters and leave the Soil Conservation Service and other USDA bureaus as institutional defenders.

There is an important two-edged sword that both aids and hinders ex/al organizations that gain political power on a temporary basis. The edge that worries economic interests is the one that builds permanent obstructions, which must be painstakingly cleared before the resources used in support of agriculture can be shifted. Any controversies that develop during this rather slow process give the existing reservoir of ex/al groups plenty of time to mobilize around the issue at hand. An attack on the legal rights of farm workers or the government supports that do much to maintain rural communities would certainly activate dormant groups that have previously cultivated and continue to understand those supportive constituents who believe that these things matter. As an alternative lobby, ex/al groups organized to raise public consciousness of social conditions and to elicit reactions to what they consider to be improper attention to the consequences of what their representatives see as predominantly economic policy. In this sense, the inclusion of new issues permanently alters the agricultural policy process each time a new agenda item gains institutional status.

The other edge of the sword damages the ex/al organization and its cause rather than those whom the staff hopes to bring under public scrutiny. Groups that lack recognition for either themselves or the issues they represent and that are otherwise limited in their organizational resources need the support of others. As Jeffrey Berry found, organizations with a public interest must often deal with issues that others will fund rather than with the ones their staff personnel may prefer.[28] Agricultural policy, with its long-standing isolation from the intervention of outsiders, presents a related problem. Ex/al groups interested in challenging the sugar program, chemical use practices, or the raising of animals must often postpone their attention to those matters and address issues that potential political supporters find important, even in coming back to protect previous gains.

Frequently, in order to gain attention from policymaking institutions that consider themselves primarily responsive to farm production problems, relationships involve more recognizable agricultural interests than other ex/al challengers to these programs. In attempting to broaden its strategy against biogenetic research, the Foundation on Economic Trends was able to strengthen its case after farm protest groups in Wisconsin attacked bovine Growth Hormone as a further cause of the farm crisis and attendant dairy surpluses. Under conditions of commodity shortage, foundation appeals would have foundered, but when oversupplies already plagued the industry, any degree of environmental risk seemed worth examining. As a result, the larger questions of biotechnical change were subjugated to problems of farmer self-interest and, as the ex/al lobbies so often charge, simple economics.

Similarly, ex/al issues that have been incorporated in recent farm bills and other major pieces of legislation have reflected the choices made by established agricultural interests about what added provisions would best win supporters in Congress while minimally affecting farm programs and farmer income. Food stamps, farm support of minimum wage bills, votes for the Consumer Protection Agency, Food for Peace, rural development, and the 1985 farm bill's conservation provisions all came about with important support from farm or agribusiness interests. Other ex/al issues, such as limitations on research and water conservation measures, have received widespread ex/al and media attention but little in the way of either the support of established interests or legislative action. It may well be that the resource limitations of ex/al groups, despite their effective adaptation, still restrict their ability to address some critical issues even when these organizations are being assisted, with some reluctance, in penetrating some open policy windows.

8

Consultants and Experts

Research on interest groups treats lobbyists as if they occupied integral positions within the organization that they represent. They need not. The vast amount of political and media attention focused on former presidential-assistant-turned-lobbyist Michael Deaver during 1986 was as much a reaction to the flaunting of his private practice's multiple clientele as to his possible abuse of White House connections. It seemed outrageous to many observers that a lobbyist, like an attorney or medical doctor, would sell his or her skills to whatever organization could pay the fees.

Public images have not yet caught up with the changing political reality of more complex representation. Lester Milbrath identified only three multiple-client lobbyists among his sample of 114 Washington representatives in 1963.[1] Apparently very few organizations worked with them. In his comprehensive 1956 analysis of who was who among private sector agricultural policy activists, Wesley McCune cited no private consultants or important lobbyists for hire.[2] As mentioned in Chapter 3, hiring consultants for various public affairs purposes is now common among agricultural interests; nearly everyone does it. Moreover, in a *National Journal* poll of Washington lobbyists, multiple-client representatives dominated the rankings of best individual lobbyists.[3]

These free-agent public affairs specialists now have far more policy responsibilities than was once imaginable. While Milbrath found both subject matter experts and those with a special talent for promoting legislation, present consultants have more varied skills. Economic analysts and public relations specialists as well as experts in public-opinion polling and fund raising also provide important services.

In addition, numerous private foundations incorporated as nonprofit research institutions supplement the array of public affairs consultants by both contracting and doing conventional policy analysis without the encumbrances of an ex/al agenda or mandate. By encouraging research and seeking the involvement of policy area specialists primarily from universities, these 501(c)(3) foundations and institutes provide useful information that routinely enters and influences the policy process. Because of the status of the institutions they represent, these experts have become participants in agricultural policy debates, even though what they propose may be narrow demands that advocate the distribution of specific program benefits.

This chapter looks at the contribution of external experts to organized interests, which increasingly provide services integral to the expansion of these interests into issues of public policy. Theirs is an ancillary service, however, one that private organizations could do without but one that private representatives find useful in organizing a sophisticated and current lobbying effort.

USING CONSULTANTS

There are at least two explanations as to why private interests retain public affairs consultants. Many respondents believe both to be partially accurate, an opinion underscored by a recent poll that ranked the same consultant lobbyist as the "best" and the "most overrated" representative in Washington.[4] One view of consultants holds that they offer the best services available, representational resources that groups and firms otherwise would be without.[5] To this extent, consultants simply offer a strategic advantage for the contractor and are brought in to work with ranking officials of the organization.

The alternative view, somewhat more cynically offered by several group representatives who hire consultants, is that multiple-client representatives have created their own market independent of whatever lobbying resources their employers already control. According to those who articulate this position, public affairs consultants thrive from being overrated in that they escalate the amount of resources an interest will direct to a specific issue simply because of their presence and their reputations. From this perspective, an otherwise well-staffed organization may be at a strategic disadvantage in many instances if it fails to hire those consultants whom policymakers expect to be part of the process. The employer may or may not gain expertise or skills that supplement those of its own public affairs staff; it does, however, gain an often important ally, albeit one that top group or firm officials must watch carefully. In other instances, cynics charge, the retaining interest gets nothing for its money.

Reconciling these contrasting views of the consultant's role, at least from what respondents say, is relatively easy if their services are viewed as an important way of avoiding risks in an uncertain political situation. Consultants, at least in agricultural policy, offer a great deal in terms of knowledge, skills, and contacts. Most organizations would find it useful to add their experience on at least some issues. For most interests, however, present lobbying resources are adequate. The difficulty of when to contract lies in the extent to which a potential consultant employer is willing to assume the risks of *not* buying some extra degree of support and security. There is no optimal answer. The more important the pending issue or problem, the more incentives exist for contracting. As a result, an organized interest gravitates to those with the best reputation for a specific type of service.

Seen from the perspective of risk avoidance, hiring consultants is not simply a cost-effective way of securing greater talent for a minimal investment. While some organizations such as the National Soybean Processors Association find it more economical to use Hauck and Associates rather than a full-time staff to represent the group, other trade associations spend more for public affairs consultants than they would for additional personnel. Several agricultural consultants command more in fees from a collection of organizations with which they choose to work than any single interest could justify in salary for these same individuals. Yet it is the individual, and the reputation that surrounds that person, that interests seek.

Since contracting organizations can be somewhat unclear as to whether their goals demand additional lobbying resources or the support of a recognizable figure in agricultural policy circles, hiring practices vary according to how comfortable organizations feel in working with consultants. Some agricultural interests seldom hire consultants. The officials of these organizations, all of whom have experience with consultants, feel that full-time employees understand the firm or group better and can be better trusted to represent it discreetly and confidentially. Cargill, for example, prefers to develop an in-house staff of everyone from tax experts to economic analysts to avoid misperceptions created by those who may not fully understand corporate needs. Consequently, consultants are given a smaller role in central public affairs by such companies.

Although many lobbyists agree that problems of organizational awareness exist regarding consultants, most of them minimize difficulties and prefer to use them regularly, allowing the retained representative considerable autonomy. The careful selection process that most organizations use and the consultant's own reputation lead to generally high satisfaction with the relationships that have developed. As one long-time consultant acknowledged, "Unreliable people seldom last long in this business. Very few people succeed in establishing sound reputations [after they begin consulting] because you really have to have [your expertise established] before starting." In general, those who rely heavily on consultants grade them highly in terms of their skills, experience, access to policymakers, policy credibility, and client loyalty. Proponents rated consultants as generally brighter, more hard-working, and more knowledgeable than association staff. "In this sense," said one trade association director, "you get what you pay for. And you pay a lot."

Problems with consultants were few. Confidentiality was usually a minor concern since, in the words of one corporate official, "You can't hide anything in this business anyway." Conflicts of interest between clients were not problematic. Both consultants and contracting organizations noted a self-selection process that generally linked compatible interests together for multiple-client representatives. Since active interests are served best when policymakers know what prominent consultants represent an organization, clandestine arrangements are rare. Neither the reputation of the consultant nor the status of the issue can be enhanced by disguising who represents whom, especially since consultants are often used to provide direct cues to decision makers over the merits of a pending decision. Even in 1985, months before Michael Deaver gained public notoriety, lobbyist respondents cited his firm as an example of how to create problems for both the representative and the clients. Deaver's was a representational approach to avoid, these Washingtonians asserted, because his well-known list of mixed clientele would eventually raise questions about the integrity of the lobbying process.

The only operational problem frequently mentioned that was associated with consultants had to do with the circumstances under which they were employed. Both contractors and contractees emphasized that the specific organizational needs and purposes for which the consultant was being hired had to be carefully determined and mutually agreed upon in advance. Too frequently, respondents claim, they are not, since reputations for personal success rather than specific service needs often determine hiring. The result is, in the words of one corporate executive, "plenty of good projects that serve no useful purpose." With careful planning and a good working relationship between consultants and organization officials, the only other problems with consultants are some suspicions and insecurity among permanent employees who perceive nonpermanent employees as threats to their jobs.

CONSULTANTS AS TECHNICIANS

Since public affairs consultants are a rather newly expanding force, there is some tendency to think of them as high-tech operatives of the computer age.[6] A high-tech description is only partially accurate, however. As mentioned in Chapter 3, consultants and their clients think of retained public affairs work as divided into three neat categories: political contact, public relations, and policy analysis. The latter two, as supposedly distinct enterprises, are steeped in an especially high-tech mystique.

In agriculture, a few well-known firms promote their technocratic images. Most of their projects are analytical, usually with an economic emphasis. As technicians, policy analysts emphasize that their potential for influencing the policy process lies in their command of details.[7] Their tools include large data bases,

quantitative methods, and computers, but these tools are of secondary importance to the analytical consultant's experience with and knowledge of the conditions, issues, and problems of agriculture.

Although other agricultural economists are involved in the business, the two
best-known agricultural policy analysis firms are Abel, Daft, and Earley[8] and the
newer Economic Perspectives, Inc. Both firms have small offices in which the
managing partners apply a hands-on approach to each project. Clients contract to
acquire market information, quantitative projects on proposed management decisions about such topics as new facilities, shifting market strategies, or economic
projections and forecast. Although the resulting analyses are sophisticated, contractor respondents indicated that the experienced judgments of a Lynn Daft or J.
B. Penn were really what they prized. Young in-house technicians can generate
data and projections competently, but their conclusions will be inadequately
grounded in what customers see as the contextual circumstances of agricultural
policy. In essence, even as technical analysts, these consultants are valued because of the conclusions they are capable of drawing after years of advisory experience in the White House, USDA, and Congress. Were it not for this perceived
value, neither firm would be at its present level of demand. As one contractor
stated, "Lynn Daft understands the impact of wheat programs. My staff, at their
best, just combine variables."

Even though public relations firms have received an inordinate amount of
press attention as "superlobbies," there is less reliance among agricultural clientele on the other type of public-affairs technocrat. One firm, Hill and Knowlton,
Inc. (the former Robert Gray and Co.), nearly dominates the public relations market in agriculture. "Hiring Bob Gray," said a commodity lobbyist, "I think, is a fad."
Nonetheless, a diverse group of clients felt it necessary to use the specialized
skills of Gray's agricultural specialists, a team brought together in 1984 in antici-
pation of opportunities available with the politics of the pending farm bill. Several
trade associations, troubled by high price supports, contracted for advertising
campaigns. At the other extreme, discontented Nebraska wheat farmers used
the firm in an attempt to call even more public attention to their already well-
promoted plight.

As the above examples indicate, public relations firms in agricultural policy
are something of a last resort for interests facing an immediate crisis or a pending
opportunity to influence policy adjustments. When the *New York Times* equated
boards of trade with Las Vegas gamblers, the Futures Industry Association (FIA)
felt that the time for a professional defense was at hand. Already under siege by
farmer protestors and criticized by congressional members, the FIA responded
by soliciting proposals from Gray and other public relations firms. The FIA eventually contracted with Daniel J. Edelman, Inc., to perform the typical, well-
refined services of developing a public opinion poll, assembling a media packet,
organizing an industry tour of in-house speakers, preparing mailings, and designing both advertising and editorial materials.

Despite these crisis-oriented examples of use and the 1984 start-up operations of Gray and Company, agricultural representatives remain far more skeptical of the technology of public relations than they are of economic analysis. Respondents believe that, to conduct their business and to lobby, they must have technical information and expert advice. Unlike other industries, these respondents do not believe that the food industry and agriculture need the image-building assistance of public relations generalists who specialize in techniques rather than in agricultural affairs except as some last resort on a particularly touchy problem, such as refuting the claims of agrarian reformers. Product sales are another matter, one in which extensive use of advertising campaigns has long been established. However, these responsibilities have been delegated to promotional boards and councils that usually work through advertising firms in New York, Chicago, and Los Angeles, rather than the policy-oriented image managers of Washington. Washington firms are more likely to be used when, as was the case with the FIA, an immediate threat is made where political problems are already acknowledged, or they are used to promote such things as value-added products when export enhancement depends on congressional action.

CONSULTANTS AS POLITICAL OPERATIVES

In his study of administrative policy analysts, Arnold J. Meltsner found a distinction between those who wanted to make policy and those who thought of themselves as technicians.[9] Bureaucratic technicians saw policymaking as directed too much toward salesmanship and preferred instead to do research with little or no concern for how it would be received. This distinction has some importance for private consultants because it helps to differentiate between the work of various free-agent entrepreneurs as they define and structure their services.

Among public-affairs consultants in agriculture, the policymaker/technician distinction is much less firmly established than for bureaucrats who exist in a more complex environment, in which a variety of policy alternatives can often be examined in some isolation. Private-sector consultants, in contrast, understand that their clients look for immediate or at least potential payoffs in the work that they contract out. On political or policy matters, where the prospects of gains and losses are real, clients want political or policy help rather than just more information for its own sake. For obvious reasons of client satisfaction, even consultant technicians cannot retreat from political responsibility just by preparing volumes of detailed information. Contractors want the expert's advice about the meaning and implications of their findings. On some occasions, this means that technical advisers assume some lobbying responsibilities.

The policymaker/technician distinction is important for determining what lobbyists and advisers as counterparts to bureaucratic analysts do for those interests that hire them. Consultants who find their skills and experiences especially suited to advisory work want to avoid a lobbyist label and the negative impressions of political hucksterism that it creates among many business executives, the consultant's most frequent clients. That label limits both their perceived status with others and their potential to provide a broader range of public affairs services. One veteran consultant who testifies regularly before Congress on behalf of clients insists that his firm does not lobby. Especially for newer technical firms, active lobbying also creates too many opportunities to antagonize possible future clientele.

Labels and related images are important primarily to establish uniqueness, because unlimited competition between consulting firms fails to reinforce the importance of the specific expertise of individual consultants and firms. Consequently, consultants try hard to emphasize what they—rather than someone else—can do for the type of client they serve. While a few exceptionally large firms such as Robert Gray's Hill and Knowlton division exist in Washington, most consultants attempt to create the same kind of specific niche in the policy process as do conventional agricultural interests.

At least in agricultural policy, consultants have been quite successful in finding their niches and gaining widespread recognition among agricultural interests for what they do for whom. The technical firms are no exception. With its ties to previous Democratic administrations, for example, Abel, Daft, and Earley claimed a natural clientele of wheat and corn interests along with related midwestern agribusiness corporations that benefited from moderate price support programs and stable supplies. Sugar interests and others that want a heavily subsidized sugar program and that have been at political odds with Martin Abel or Lynn Daft in the past look elsewhere for analysts. Even at Hill and Knowlton, the announced emphasis is on its public-relations capacity rather than on its "super-lobbyist" press image, and clients work through experienced associates with whom they share a political background rather than just any designated subject matter specialist. Hill and Knowlton associates, when involved with complex policy problems, find it necessary to contract out for analytical services with firms such as Economic Perspectives and to make referrals to more knowledgeable agricultural policy specialists from outside the company. Even the biggest firms understand that their resources, when compared with others, fail to allow them to be all things for all clients.

A small number of consultants, offering more political than technical services, occupy other prominent advisory niches in agricultural policy. The management firm of Food Systems Associates operated by Tom Veblen does no lobbying but proselytizes on the economic virtue of its clients' developing greater political astuteness. Most of the others do some technical analysis, diligently watch pending political issues, interpret political events, work with informed groups of officials, and advise clients on strategy and tactics. Three are especially well known. John Schnittker and Martin Sorkin, both now semiretired and with their own small lists of clients from established agricultural interests, are primarily domestic policy specialists. Dale Hathaway of Consultants International is their counterpart on trade and international matters. Though not all three register as lobbyists, all are active and highly visible among agricultural officials and other Washington representatives.

The remainder of agricultural consultants willingly identify as lobbyists. However, many prefer to use the term *issue manager,* implying a more comprehensive enterprise than merely making political contacts. These consultants— many of whom are attorneys—watchdog, pursue strategies with their clients, make and help make political contacts, raise and distribute campaign funds, and litigate. In essence, they are not so different from those who have achieved adviser status, except that lobbyist consultants tend to be less actively involved with internal management affairs. On the whole, lobbyist consultants also make greater use of junior staff personnel than do analysts and advisers, but, as with other multiclient representatives, client expectations and the specialized demands of policy problems mandate extensive personal involvement on major issues even for lobbyists. In larger firms, minor issues and most of the monitoring gets relegated to junior employees who frequently use these opportunities to establish their own reputations as agricultural experts. The extensive use of junior employees frees many senior consultants to engage in brokering relationships between clients and policymakers in order to secure the recognition that many interests feel is essential and otherwise absent. Some of the better-known lobbyist consultants have made most of their reputations as fund raisers who are able to deliver campaign dollars above and beyond PAC contributions. Several are known by legislative staffers as "bagmen," lacking substantive knowledge of client needs but still able to get in to see their bosses.

Although a number of firms are always attempting to develop agricultural connections, only about a dozen regularly though not always exclusively represent food, fiber, and farm interests. Most of these have even narrower reputations for whom they represent within agriculture than do the technical consultants. For example, Jaenke and Associates is identified by long-standing ties to the cooperatives. Heron, Burchette, Ruckert, and Rothwell represents fruit and specialty crop interests and their trade and marketing order concerns in particular. Other California interests, especially large farms, provide the focus of Wesley McAdam. Patton, Boggs, and Blow is known for its commodity user clientele. Smaller trade associations do considerable business with Hauck and Associates. Sam White, Larry Meyer, and a small number of other lobbyists loosely affiliated in two Washington offices represent most small producer groups that come to town with farm legislation problems. As part of their co-op business Martin Haley Companies, after wooing the American Agriculture Movement, became increasingly identified with agricultural alcohol interests and alternative energy sources.

The degree to which these and other firms represent most agricultural interests that require policy assistance leaves little room for agricultural policy newcomers to attract clientele. Only rarely does an exception occur. For example, Washington real-estate attorney Stephen Huttler successfully represented Farmers for Fairness after a chance meeting with group leaders in Georgia. Even with his success, however, a lack of recognition among agricultural interests made it difficult to secure additional producer clients. Richard Lyng, now secretary of agriculture, and William Lesher had no such problems with their new firm after these Reagan appointees left their ranking USDA positions in the midst of preparations for the 1985 farm bill. While they attracted numerous clientele on the basis of the certainty of their political access, Lyng and Lesher soon gained a reputation as useful allies to employ in opposing price support programs. Robert Gray also benefited from close Reagan ties. His firm attracted clients because their principle service, public relations, was unique to agriculture and because the firm was reputedly a Washington success. Nevertheless, most clients of Lyng, Lesher, and Gray retained their ties with old advisers, simply adding to their lists of multiclient allies.

Firms such as the former Lyng and Lesher not only represent compatible interests; they also provide another service by helping bring these interests together. As long as E. A. Jaenke represented multiple large midwestern cooperatives such as Farmland Industries and Land O'Lakes, the co-ops possessed a common organizational spokesperson who addressed their related interests both within the co-op structure and without. Their policy positions were strengthened somewhat in Washington as a result of a widespread belief that the co-ops stood together.[10] Variations on this representational advantage are also served by consultant lobbyists. Some lobbyists have been able to put together coalitions based on their own multiple clients. Marshall Matz, representing both the American School Food Service Association and some food suppliers, gained his peers' respect for being able to get clients to see their shared interests and work together in developing both policy positions and business relationships.

Such cooperation is not restricted to the clients of a single firm or lobbyist, however. While not intentionally integrated in any way, the community of agricultural policy consultants is small enough for its practitioners to know one another's business and capabilities. As a result, working together across normal lines of expertise is commonplace when the need arises. E. A. Jaenke, serving as issue manager for the Farm Credit Council's proposed 1985 credit assistance bill, assembled a team of lobbyists and analysts from several consulting firms as well as from the Farm Credit System. By carefully considering the potential contributions of other consultants, Jaenke was able to bring together what he considered good people with bill drafting skills, banking expertise, White House access, USDA access, and experience with both parties in Congress. By doing so, he became a successful lobbyist-as-manager, whose resources not only were greater than any competing interests but who also confronted nearly every conceivable political and technical obstacle in the policy process. Ultimately, this unmatched degree of lobbying and policy expertise is precisely what the consulting community has been designed to bring about by its few dozen entrepreneurial founders.

WHO HIRES CONSULTANTS?

Harold D. Guither's study of food lobbyists, although failing to examine the role of consultants, finds sugar interests actively represented by such individuals and their firms.[11] But Guither rarely mentions consultant lobbyists representing other interests. The time frame could be considered an important reference, since Guither's study was done in conjunction with the 1977 farm bill, when political battles over sugar programs and quotas were at a peak. As indicated by the 1985–1986 respondents for the study presented here, and as frequently noted throughout this chapter, consultants are likely to be attractive to interests in political trouble.

Information provided by lobbyist respondents indicates that the use of public affairs consultants is not triggered just by an organization's escalating political involvement. That is to say, consultants have not become a necessary political tool in a changing political environment. As Kay Lehman Schlozman and John T. Tierney discovered in their study of Washington lobbyists, there have been no representational revolutions in strategy, despite the advent of new and more sophisticated techniques.[12]

Their widespread use notwithstanding, agricultural policy consultants appear neither to be indispensable nor to have changed the style of interest representation. To conclude that they offer some dramatic contribution to the representational process would make too much of consultants' expertise, respect, and status and of the widespread use of consultants by especially active interests. The consensus of respondents was best summarized by a consultant who suggested that he may occasionally provide a winning margin, "an edge."

Who uses the edge? Producer and producer/middlemen interests, sometimes even those with a confrontational or protest history, are frequent contractors. While those producers—such as sugar and peanut growers—with particularly difficult political problems rely heavily on consultant lobbyists to protect threatened programs, other equally threatened commodities do not. Dairy interests that want to protect their support programs have not been inclined to use consultants, even though their well-organized opponents have done so extensively.

Another problem with associating the use of political consultants with the difficulty of the problems the organization faces has to do with measuring the degree of any political threat. There is no valid basis of comparison, either for the interest group or for the observer. For example, even though wheat programs were not especially threatened by a loss of supporters during the 1985 farm bill deliberations, wheat interests found themselves in great conflict over mandatory production controls. No one, with the exception of after-hours-barstool-analysts, found it worthwhile to determine whether the National Association of Wheat Growers, the National Milk Producers Federation, the American Sugar Beet Growers Association, or the Peanut Growers Group faced the most serious political problems with the 1985 act. From the perspective of minimizing policy relevant risks, the situation for each of them merited using whatever lobbying resources the group could bring together. In fact, some of these groups used consultants, but none used them in the same way. It can only be concluded that producer interests use consultants under highly situational policy circumstances dependent upon some more pressing intra-organizational decisions made at the top. These decisions may take any number of factors into account, ranging from talents available within the group to concerns about which specific policymakers need to be reached. Small commodity organizations with single-project interests and small staffs rely on consultants most heavily, but as individual organizations are no more likely to employ them than are large multipurpose groups.

The common denominator alluded to by producer representatives in their assessment of the need for consultant advice dealt with the organization's perceived need to better enter the mainstream of agricultural policymaking. To many respondents, the well-known consultants were seen as the ultimate agricultural insiders because of both their background and their contacts. In the infrequent instances in which agrarian protest activists sought to use consultants, interests such as the American Agriculture Movement and Nebraska Wheat Board were motivated by a desire to become, at least temporarily, partners to the policy process rather than the confrontational opponents they had become. In both cases, protest group activists were willing to encounter internal conflict within the protest movement in order to move into a position from which they felt able to negotiate specific policies with government officials. Consulting firms such as those of Martin Haley and Gray were seen and sold as the only available vehicles for such a transformation.

Nonprotest groups are attracted to consultants for the same reason, even though their specific public affairs needs may vary. Many dry-bean growers, believing that their prices would be hurt as a result of oversupplies threatened by 1985 farm bill provisions, have argued that a retained Washington lobbyist is essential in order to avoid future policy surprises. American Farm Bureau Federation staff members who favor the creation of a Political Action Committee believe that an experienced consultant should be brought in to structure PAC operations. In these and other instances, producer interests want only to be brought closer to policymakers than their own in-house resources will currently take them. Although attorney Julian Heron continually advises new clients that his firm provides no services that could not be done by group members, California producer/middleman organizations have felt it almost mandatory to contact Heron on matters of marketing orders and foreign trade. When an organization's

activists and representatives feel sufficiently left out, or are accused of misman-
agement, the consultant's wares become attractive because feelings of uncertainty are so great. Experience, skills, contacts, issue management, economic
analysis, public relations, credibility, or some combination of these characteristics appears to be a highly desirable means of moving to the center of that political
arena in which so many interests are competing for attention. Heron's clients, as
one expressed it, "feel better protected with him on our side." Such attitudes as to
what actual policy gains come from the use of consultants almost certainly account for the competing perceptions about their strategic importance. Are agricultural experts unraveling the policy maze or just making it so complicated that
someone feels a need to include them in it?

This perception of an organization's feeling left out of a complex agricultural policy process helps to explain why agribusiness industries and their chief
executives rather than producers are the consultants' most frequent clients. As
seen in Chapter 6, agribusinesses generally have been among the least active
agricultural interests. Consultants indicate that agribusiness as a whole has neglected agricultural policy, been willing to live with the previous costs imposed
upon it by policy decisions, and only now wants to begin to influence the process. "But most firms need to walk before they can run, understand what's going
on before they tell people what to do," as one consultant concluded. Consultants
uniformly portrayed the typical agribusiness executive as highly uninformed
and frequently naive about agricultural policy, its implications, and the manner
in which it develops.

The increasing realization by corporate executives that agricultural policy
has important consequences for their firms provides a ready consulting market
for agricultural experts from inside Washington and from universities. For most
consultants, the majority of agribusiness contracts provide "very simple services," such as periodic informational seminars and meetings on the one hand and
updated interpretations of policy and policy events on the other. Consultant attorneys also do considerable business with these same clients by outlining what
firms need to do to conform to policy decisions and by evaluating whether corporate responses to legislative and administrative rulings are adequate. Attorneys
also watch trade association procedures to avoid antitrust violations that could
result from cooperative actions between business competitors. Since much of
this information can be complex and technical, many agribusiness interests may
be referred to and use the unique services of several consultants even without
getting involved in projects that require lobbying.

Simple services, as several consultants call them, are not restricted to either
new agribusiness clients or firms just beginning to show an agricultural policy
concern. Most agricultural consultants have long-term clients they continue to
inform and advise as little more than policy watchdogs. As a result, many of these
firms employ no internal public affairs staff of their own but rely on periodic visits
between consultants and business managers to guide business decisions. Beyond

this political counsel, several corporations restrict their lobbying to back-home contacts with their own legislators. When a business problem or some advice results in an actual lobbying effort in Washington, most consultants will make a few calls or visits. Any extensive lobbying on behalf of even the oldest client will result in referral to consultants who specialize in political contact work.

The use of retained consultants, sometimes as many as four or five per company, does lend agribusiness considerably more Washington presence than would otherwise be suggested.[13] Because agribusinesses use consultants with greater frequency than do producer interests—"Tenfold, probably more," suggests a veteran multiple-client representative—policymakers frequently identify this visible set of participants as extensions of the agribusiness community. As consultants testify, seek information from government officials, and become sources of information for the media and policymakers, the firms they are thought to represent gain reputations for shrewdness and political acumen. One congressman observed, "Quite frankly, when I see Dale Hathaway or John Schnittker, I immediately think of certain corporations. I admire these guys and what I assume to be a corporate position that one or the other has helped create benefits because of my making the association."

Despite heavy producer and agribusiness use of consultants, no one controls this lobbying resource. Most of the public-affairs consultants with agricultural policy reputations have either represented or done work for foreign governments and businesses as well.

These consultants generally gain a more favorable reception among policymakers when they present the probable international consequences of U.S. policy decisions. When a foreign leader such as South African Foreign Minister Roelof Botha directly lobbies against U.S. economic sanctions by threatening a grain embargo, policymakers lash out at foreign interventionists. A consultant can carry similar messages by presenting, as a knowledgeable observer, probable international reactions even when such scenarios are merely being transmitted by the consultant as an intermediary. Many consultants have also advised such single-project agricultural interests as labor and transportation, where organizations are lacking in knowledge about agricultural policy but capable of paying consulting fees. A few ex/al organizations, such as Public Voice, use these same agricultural experts. Often, as with foreign clients, consultants are really serving as intermediaries between groups that perceive themselves as disadvantaged by their outsider status with those policymakers whom they feel reluctant to contact on their own.

The ex/al organizations, often beset with financial problems and involved even more selectively than most other interests in agricultural issues, generally find agricultural consultants of little use. Ex/al staff, in fact, tend not to trust even the most liberal of them because these consultants have for so long been part of the farm and food policy process. While this attitude further restricts a set of organizations that already suffer from a lack of knowledge about both policy and process, ex/al groups are able to establish other relationships that provide them similarly helpful and no less expert allies. Those organizations both employ and interact with a cadre of recognizable ex/al consultants.

As discussed in Chapter 7, ex/al organizations tend to reciprocate and support one another within issue areas. This advocacy "networking," as activists like to call it, in which like-minded organizations join together to enact or protect public policies, remains only one form of mutual support. Environmental or animal rights groups, for example, also engage in "exchange networking" to share money, office space, supplies, and expertise.[14] When agricultural issues achieve an ex/al group's priority status, exchanges flow within and across issue boundaries. Staff members, in return for whatever renumeration is possible from grants and contracts, frequently serve as consultants for one group while employed by another. If funds are unavailable, interorganizational goodwill is usually sufficient to encourage the transfer of whatever information various ex/al representatives possess about agriculture. Within Washington, when agricultural issues are at hand, there are few isolates among the ex/al groups because their representatives believe that a united front is necessary to overcome what they see as the intransigence of agriculturalists.

Much of this cooperation characterizes relationships between ex/al group activists and many of the Washington consultants who work in related issue areas. Using their experience, skills, and contacts, ex/al consulting firms have become extensively involved in providing services to trade associations and business firms that must deal with an ever increasing number of regulations and laws as well as a public sentiment aware and supportive of ex/al concerns.

While many of these consulting firms' fees come from interests to which most ex/al lobbyists do not feel close, consultants nonetheless maintain ties to advocacy organizations. Such linkages are important if for no other reason than the consultant's need to keep up with all facets of an issue area. For most, however, shared experiences in the past and common policy concerns in the present provide even better reasons for remaining involved with the advocacy groups. In this sense, both agricultural and ex/al consultants are probably characteristic of some broader yet unresearched set of private-interest representatives that exist in most if not all issue areas of U.S. public policy.

AGRICULTURAL EXPERTS
AND RESEARCH FOUNDATIONS

A small number of private research-oriented institutions, or think tanks, exist where staff publicly rule out lobbying as one of their purposes. Yet these foundations are among the most recognizable and instrumental components of the agricultural policy process; the ideas they generate often have great influence on the demands and proposals of those who want to shape legislation. Other factors contribute to this paradox in which prominent organizations both want and deny their direct influence. Like other 501(c)(3) foundations, the staff of these organizations involve themselves with policymakers. Unlike their counterparts who share an ex/al approach to public policy reform, however, these foundation representatives are seldom avowed advocates of specific proposals. They prefer careful presentations of the pros and cons of various policies or proposals. Nonetheless, both respondents who like them and those who dislike them see their work as predominantly shaped by the perspective of those tested and established as agriculturalists rather than as dissidents.

Adding further to the paradox is the degree to which these private interests are akin to agricultural consultants. Their work is essentially alike, in that it contains a technical emphasis with a strong recognition that policy decisions are made in a political universe that must be consciously shaped. The foundations focus more on identifying problems and developing operational proposals from such conceptual ideas as free trade or agricultural sustainability, but they are frequently criticized for their inability to package their solutions politically. They are client directed, however, in that the foundations frequently focus the attention of conferences and research on topics of specific concern to the corporations, organized groups, and government agencies that contribute to their operating budgets. Furthermore, like the consulting firms, the foundations gain their credibility on the basis of their agricultural expertise. Experience, background, and skill help to define expertise.

Finally, the paradox is completed by the degree to which consultants are involved in major foundation projects. Each of the private foundations that held pre-farm-bill conferences in 1984 and 1985 used prominent agricultural consultants from the select ranks of public affairs advisers. Each of these individuals had periodically worked with the foundations before, either as a consultant or in a previous policymaking position.

Consultants do not sustain the research foundations by any means. While a few in-house fellows are employed, most projects are completed by other independent contractors and grant recipients with temporary responsibilities to the organization. The large number of agricultural experts affiliated with land grant colleges provides a reservoir of academics who have superior qualifications and the organizational means to undertake short-term assignments. They not only enhance the reputations of the foundations as sources of independent policy ideas; they also bring the foundations into contact with an even broader array of academic experts who work with grant recipients. Project reports and recommendations, as a result, are distributed widely among the academic community of agricultural-policy experts, a set of individuals influential in generating ideas as to appropriate policy directions.[15]

Despite such operational generalities and the small number of organizations they characterize, there is no single description that best fits the foundations involved in agricultural policy. And they are certainly not agricultural equivalents of the RAND Corporation, a large defense-oriented scientific adviser to the military on planning, operations, and weapons-development policies.[16] The organization that most approximates RAND projects is Resources for the Future (RFF), an organization with several governmental research contracts on soil and water conservation. RFF is becoming better known within agriculture for more traditional policy concerns through its National Center for Food and Agricultural Policy. The center, established by a multimillion-dollar W. K. Kellogg Foundation grant in 1984, emphasizes leadership development and the communication of policy-relevant information as well as analytical projects. These responsibilities give the center ongoing opportunities for involving policy experts with policymakers.

The Illinois-based Farm Foundation is equally immersed in agricultural policy debates and has been since 1933. However, despite its policy involvement, the Farm Foundation has neither the analytical capacity, the project funding, nor the international concerns of RFF's National Center. It remains an important forum for presenting and discussing policy analysis. The Giannini Foundation of Agricultural Economics at the University of California is beginning to develop a somewhat similar reputation with, perhaps, more of a technical emphasis.

Within Washington, other research foundations periodically fund agricultural policy projects that have prescriptive intentions. The American Enterprise Institute (AEI) and the Brookings Institution have long-standing reputations for both undertaking analytical projects and using them for timely and constructive policy recommendations. The International Food Policy Research Institute lacks the prestige of the AEI or Brookings but is beginning to take similar advisory stands on U.S. agricultural policy. On a more ideological basis, the liberal Center for National Policy and the conservative Heritage Foundation have taken to issuing advisory reports on agricultural issues. Both do well-substantiated work but lack the detailed analysis and original research characteristic of RFF's more extensive but less frequent studies.

The advent of widespread attention to the financial crisis in agriculture in the 1980s attracted considerably more foundation involvement in food, farm, and trade issues than previously. Foundations such as the Roosevelt Center for American Policy Studies suddenly sponsored a Food and Agriculture Policy Project. The Heritage Foundation began issuing agricultural reports only by redirecting the efforts of its regulatory specialists. Whether this extensive involvement will continue is uncertain. Even if it does not, organizations such as Resources for the Future's National Center, the Farm Foundation, and Brookings are sufficiently endowed and historically interested enough in agricultural policy to continue mobilizing experts who can undertake relatively independent research projects. More importantly, the attention these policy analysis foundations command will continue to make their conferences and publications important forums for disseminating the expert advice of Washington consultants and, especially, non-Washington policy experts. In that sense, the expectations of other policy participants will continue to keep a few select foundations identified as influential private interests even if they fail to lobby, in a traditional sense, on behalf of the proposals their experts develop.

A FINAL COMMENT

Consultants and foundations have come to play an especially important and permanent role in shaping the focus of agricultural policy debates. Theirs is a multi-dimensional role, ranging from providing political access to developing policy information. The importance of these experts is often exaggerated. Nonetheless, both lobbyist and policymaker respondents agree that experts as policy players immeasurably strengthen the representation of at least some interests and ideas as these are brought to the Washington policy arena. In the 1980s, that strength has been most effectively used as a counterpoint to grassroots agrarian activism.

9

Coalition Politics:
Dilemmas and Choices

The last five chapters have examined a large and diverse array of private interests that represent various dimensions of American agriculture. Relative to positions on public policy, each organized interest can best be understood as representing a unique niche in either the substantive or procedural realm of policymaking. Not all organizations are continuously in politics, and there is some minor overlap between a few groups that wish to represent the same collectivity of firms or farmers. Moreover, there are no obvious issues and proposals on the agricultural policy agenda that some private interests do not represent.

Yet there is little sense of a collective system of representation in American agriculture. Although each set of private interests—producers, businesses, ex/al groups—is characterized by ongoing interorganizational policy relationships, common goals do far less to structure this interaction than do single organizational needs. Observers must be struck by the degree to which private interests of all sorts carry out their work in cooperation with others, especially as ex/al interests have crossed over to work in conjunction with agriculturalists. It has been impossible to examine the representational involvement of any type of private interest, from the broadly reformist to the single-project player, without referring to which organizations work together, support one another, or share resources in attempting to influence the policy process. Paradoxically, however, all of these relationships have been observed to be ones of mutual cooperation in the pursuit of policy goals of unique interest to each participant rather than any arrangements built on the development of a comprehensive agricultural policy or even comprehensive review of a policy problem such as pesticide use or farm income. Repeated reference has been made to interorganizational relationships that deal only with bits and pieces of current policy problems. Not a single respondent indicated that cooperation between organized interests was intended to be anything more than a tactical means of dealing with the conventional problems of lobbying and keeping organizations operable. Only the continuing prominence of such major issues as price supports and export enhancement and the necessarily ongoing advocacy of many of the same interests on these issues provides any reason for thinking about interest representation from a systemic as opposed to a highly group-specific perspective.

The commitment of individual private interests to their own organizationally determined goals becomes especially apparent in examining the coalition behavior of groups and firms. Because of the tactical orientation of organizational representatives, they view coalitions as useful in enhancing their chances of winning on microlevel policy matters. Coalitions of multiple interests have not been important vehicles for resolving macrolevel—or even mid-range—policy differences that divide agriculture. That is to say, for example, encompassing coalitions of competing interests do not come together to address the means of maintaining farm income or the degree to which American agriculture should remain competitive internationally.

This tactical, or short-term and utilitarian, view of coalitions of organized interests should not be surprising. Political scientists are not used even to addressing interest group politics and coalition building in the same context. In both theory and practice, there has always been considerable incongruity, just as there is in the reformist climate of the 1980s when policymakers ask special interests to put aside organizational problems and focus comprehensively on the problems of agricultural policy. According to the conventions of both politics and academic analysis, private interests are by nature organized around specifically uncommon desires even when their policy goals are intended to allocate economic and social benefits beyond a selective clientele to a broader polity. In contrast, coalitions are working relationships in which those with some common interest act together to resolve any serious problems that divide partners. In linking private desires to a common interest within a coalition, interest group participants must expect to forgo some part of what they organized to accomplish.

Other than a few repeated references to like-minded interests working together to share resources, group theorists have largely ignored coalition building in developing the literature. James Q. Wilson, fully aware of the inconsistencies between special interests and common purposes, was one of the first to feel the necessity of discussing the cooperation and coordination involved in intergroup lobbying.[1] Despite identifying considerable incentives for cooperative action, Wilson concluded that coalition building was a difficult—and inadequately understood—process that few interests did well. It was not a process Wilson could ignore at the time he was writing, however, because group representatives of the early 1970s were treating coalition formation as a most important strategy. Conceptually suspect or not, Wilson's major updating of group theory could hardly have ignored an organizational format that many interest representatives were refining in hopes of keeping certain issues alive politically.

Coalition politics, for example, was used by such groups as those organized around the various local government service needs of low-income residents. Representatives of these interests felt compelled to work together if they were to gain the same degree of political attention as more established and better-funded interests enjoyed. One study of organized city interests completed during that period found much of the internal management of the large municipal interests directed toward the establishment of a coalition housed within the joint offices of the National League of Cities and the U.S. Conference of Mayors.[2] That coalition sought to bring smaller associations of municipal officials together on public policy matters as well as to establish a public policy forum for integrating the political demands of cities with those from counties and state governments. Similar attempts were going on within other policy sectors. Within just a few years, leading journalist observers of interest group politics were writing that this sort of coalition activity was transforming lobbying.[3] Public policymakers, it seems, were placed in difficult situations when forced to say no to so many.

Recent studies of interest groups have confirmed the conclusion that coalitions are increasingly used as if they were important to those who represent private organizations. Kay Lehman Schlozman and John T. Tierney's survey found that 68 percent of Washington lobbyists engaged other interests in multiple-group coalitions.[4] Robert H. Salisbury and his colleagues, while studiously avoiding the term *coalitions,* found that most lobbyist respondents—including those in agriculture—named regular cooperators from other organizations with whom they worked time and again.[5] Among those surveyed for this study, every Washington-based interest representative had participated in one or more, and usually several, cooperative lobbying efforts that they identified as coalitions. Only a small number of state and regional representatives who consciously avoided Washington politics ignored multiple-group coalition politics on national issues.

Has there been a change in lobbying that has actually transformed the process of representing private interests? Any reconciliation of the discrepancies between private interests and common purposes most certainly would have created a revolution in lobbying style. Neither other recent studies nor this one identifies any such adaptability on the part of organizational representatives of the 1980s. Lobbyists simply seem to interact more often with one another in a supportive capacity because, as they see it, the total policy process is now more complex. As a result there are few opportunities to attain even narrow policy goals without cooperating with others.

Some of that increase in interaction is a matter of observation and conceptual semantics. Wilson, with his organizational emphasis, insisted that coalitions be defined in terms of "enduring arrangements" in which policy agreements were made rather than as mere coordinating mechanisms for intergroup coordination and technical assistance.[6] Even through the 1970s, this definition meant that less than a handful of coalitions within agriculture formed separate organizational structures, met regularly, elected officers, drafted compromise policy positions, and promoted their agreements to gain a joint political advantage.

Lobbyist respondents do not agree with any such rigorous organizational definition of what they consider to be multiple-interest coalitions. It would be asking too much to expect interest representatives to favor Wilson's organizations of organizations and the entangling commitments that would limit each participant's flexibility in lobbying. More than twenty veteran agricultural lobbyists claimed that they were partners in formidable, but informal, coalition arrangements during the 1960s and 1970s. These commonplace alliances did not attempt to order the relationships among participating groups; and they were intentionally short-lived, usually forming around a single and resolvable policy goal. In that sense, respondents agreed with William H. Riker rather than Wilson in that they defined coalitions in terms of the requisites of winning rather than of organizing.[7] They wanted to organize, no matter how resourceful their own interest, only to the extent that it was the least costly in time, energy, effort, money, commitment, and their own policy positions. As much as possible, group representatives sought to protect the integrity of their private interests and goals while working together. As long as winning was possible, informal coalition relationships were preferred strongly to any formal agreements if it were at all necessary for groups to work together.

These attitudes, advocating the least amount of involvement possible with coalition partners, further underscore the episodic involvement of lobbyists as they get involved within the policy process. Salisbury and his colleagues found that although lobbyists viewed interorganizational relationships as relatively unstable, they perceived policy issues to be long-lived on the political agenda.[8] When organized interests have such careful watchdogs, why do coalitions disappear, even though the issue remains with the partners? According to agricultural lobbyists, issues as political problems may be long-lived, but as specific proposals they occupy an immediate position on the calendar of policy events. The proposals, not the encompassing issues, serve as the basis for most coalitions and for most participants who want to expend as few resources as possible. As a consequence, lobbyists meet just prior to and during the time that decision makers meet. Then, bowing to political expediency and the large number of other organizational concerns and policy proposals under active consideration, participants generally move on to something else. Should a proposal be resurrected, the coalition can be reactivated.

COALITION BENEFITS
AND THEIR ORGANIZATIONAL FORMAT

Coalitions of agricultural interests are organized in a variety of ways and for different purposes, depending on how participants view their organization's service and policy needs. There have been no hard and fast rules as to how coalition supporters should structure their relationships or behave as partners. Experiences, however, have convinced coalitions that two dimensions of organizational behavior must be determined early: Will the coalition format be one of formal or informal alliance? And, will the coalition serve the informational needs of the partners or be directed toward a compromised policy position? Although lobbyists generally prefer informality, the modern agricultural lobby has at least some proponents who favor the formal policy model, the informal information model, and the informal policy model of multiple-group coalitions.

FORMAL POLICY COALITIONS

Given the reluctance of private-interest representatives to expend the costs of working toward a common purpose, it should not be surprising that formal policy coalitions in agriculture were the most recent of the three models to develop. Only a few formal coalitions have gained any sustained support from participants, and in none of these cases have their relationships lasted. It would be accurate to describe each of them as short-term solutions reached in the face of well-organized opposition from what involved representatives saw as directly competitive and politically entrenched interests. In the words of one long-time coalition activist, "You do something like this to bring order out of the chaos, not because you know it will work."

Formal coalitions of national agriculture interests can be traced first to the twenty-seven farm groups that joined the National Conference of Commodity Organizations (NCCO) in 1957. Under the leadership of Secretary of Agriculture Ezra Taft Benson and with the active support of many budget-conscious agribusiness interests, New Deal farm income maintenance programs were being threatened after a divided Congress replaced fixed price supports with flexible payments that were intended to drop as commodity supplies increased.[9] That 1954 legislation and its restrictive payments did not halt the growth of commodity surpluses, however, so an alternative approach, the Soil Bank, with its cutbacks in production, found congressional favor in 1956. Agribusinesses became even more actively hostile and mutually cooperative in their opposition, since mandatory production controls for supported crops seriously affected farm input sales and threatened to increase the price other interests paid for those restricted commodities. With considerable sympathy from the White House, and with surpluses still increasing under the Soil Bank, agribusiness appeared to be a strong policy force as yet another farm bill was pending for 1958. This strength seemed likely to increase as the stalemate in policy preferences between the partisan Farm Bureau and the NFU continued to split Congress. Perhaps, most farm lobbyists feared, conservatives would find the timing appropriate to eliminate both high-cost price support programs and acreage set-asides. It was in that context the NCCO was created.

The NCCO coordinated a lobbying strategy, mobilized farmers and most of the regionally organized farm groups, won considerable support for its cooperative tactics in Congress, and was instrumental in passing more lucrative support programs. But just as policymakers were praising the coalition for successfully demonstrating where victorious farmers had actually stood on the issues, partner groups began to break up in conflicts about who would direct the coalition and under what rules. Many representatives became convinced that, in the absence of specific policy initiatives, some partners would use the organization to grab power, using the influence and status of leadership positions and lobbying resources that everyone had invested in the coalition. In short, they feared the emergence of what might become a truly peak association representing most farmers' commodity-specific policy concerns. Such an organization was feared as much more likely to dominate agricultural policy than were either the ideological or fraternal farm groups.

Following the collapse of the NCCO, several former participants almost immediately began to hold meetings in the hope of restructuring a continuing alliance. These activists, troubled by increasing commodity surpluses and the criticisms they engendered, felt that only negotiated and collectively supported policy demands could keep farm price support issues before a Congress in which the number of rural legislators was in sharp decline. Over the next decade, the Midcontinent Farmers Association and the National Grange were most instrumental in designing and holding together an organization in which meetings could be held, but in which dominance by any interest was unlikely. Calling itself the Farm Coalition, this loosely organized group negotiated policy positions on commodity price support levels and related production programs. Between thirty and thirty-seven groups participated for each of the farm bills of the 1960s.

In 1970, political conditions once again changed the context in which farm bills were decided. A new secretary of agriculture, Clifford M. Hardin, proposed a more market-oriented approach to farm policy with less costly price support legislation; and he immediately began to meet with various agricultural interests to win their endorsement. To ensure a greater degree of Farm Coalition commitment from participants and avoid defectors in the face of these new threats, the organization was recast in 1970. Renamed the National Farm Coalition (NFC), a formal organizational structure was advanced once again. Membership was defined to prevent confusion over the coalition's identity. Officers were elected to promote leadership and designate official spokespersons. Bylaws and constitution were rejected only when the problems of the NCCO were reopened.

The National Farm Coalition became plagued with problems almost as soon as it was renamed and discussions about formal rules began.[10] After years of comfortably agreeing to compromises in the 1960s, the NFC found that even its past successes elicited criticism from most group partners facing the possible restrictions of a formal coalition. Partner groups that were credited with gaining most of their price support goals, that served as important intermediaries in negotiating price support levels with nonmember groups, and that maintained a solidarity on several successful trade agreements also criticized the coalition for obscuring their own legitimate policy demands. Their representatives also expressed the somewhat contradictory view that the Farm Coalition had consistently failed to develop policy positions on major issues in which one-for-one tradeoffs, or exchanges, were not the basis for mutual agreement, as was the case for price support levels. The defeat of the 1963 wheat referendum was cited as a case in which coalition partners failed to lend a hand to the affected interest even when the proposal set an important precedent for all of agriculture. A lack of support for cooperative marketing practices and a reluctance to address collective bargaining issues also were noted as issues in which the Farm Coalition failed to live up to its name. After Secretary Hardin's reforms made little headway in the 1970 farm bill, the NFC—even with its formal status and the continued threats of a free-market approach to the 1973 farm bill—retreated to a position in which members met only for the purpose of negotiating joint but nonbinding commodity proposals on future farm bills. Proponents of this arrangement argued that these were the only public policies in which each interest had enough stake to negotiate a common position. As a formal organization after 1970, the NFC was involved in fewer issues than was the informal alliance of the 1960s.

Conditions were not much different for the formal coalition of thirty conservation groups that sometimes expressed their collective interest in agricultural policy, the National Resources Council. Although some members, with council support, were able to take the initiative in the 1950s and 1960s to redirect land set aside under the Soil Bank for wildlife habitat, the council received mixed messages from partner groups about further challenging agricultural issues and interests. Although participants, such as representatives of the Izaak Walton League, believed that it had been essential to have a coalition of diverse conservation interests with large numbers of members fighting for farm bill recognition, there were limits to what they were willing to do. Given continuing agricultural surpluses and a growing public use of recreational facilities, many council partners wanted to push for further conversion of agricultural acreage and restrictions on farm and ranch use of governmentally owned lands. Hunting and fishing organizations, aware of the need for producer support in using privately owned converted acres and farm ponds constructed with soil conservation funding, did not want to continue to antagonize established agricultural interests. Like the National Farm Coalition, the council retreated into a position of minimal policy

involvement. Again, as had been the case with the Farm Coalition, council participants raised the paradoxical question of why they should spend time maintaining
an organization that did little to advance their private policy preferences while actually threatening each group with other policy losses. At least in agricultural policy, the council was dead soon after some partners gained the narrow policy
results that they had bargained for within the coalition.

The pattern of these three formal attempts at coalition was clearly one of
partners' feeling the need to assume the risk of collective action, resolving the
issue to the satisfaction of some participants, splitting ranks, and then running
from or at least minimizing coalition responsibilities. Lobbyists can hardly be
blamed for finding formal coalitions politically suspect, especially in light of the
costs of maintaining these alliances. In this sense, formal coalitions remind many
representatives of the trade associations that, as one disgruntled member representative claimed, "consistently address only the most easily defined common
policy denominator that is of interest to member firms." Like trade associations,
formal multiple-group coalitions seem destined only to disappoint their strongest
policy proponents because what will be selected as targeted priorities will be the
least rather than greatest number of issues possible. The unity, enforced commitment to policy goals, and collective discipline that contemporary advocates of
multiple-group coalitions value are nearly impossible to maintain for long.

The experiences of two formal coalitions of the mid-1980s reinforce that
conclusion even though no resolutions of their concerns have yet developed. Despite their partner group and policy goal differences, both have organized under
the same pressures and faced the same problems. The Sugar Users Group and
the National Save the Family Farm Coalition see policymakers deferring only to
established farm interests on farm income maintenance programs and are especially bitter about it. The passage of high levels of sugar and dairy supports and
the rejection of amendments identified with the proposed Farm Policy Reform
Act in the 1985 farm bill reinforced those attitudes. To combat established opponents, organizers of the two coalitions argued for formal organization. They
hoped that collective action would keep at least some partner groups continually
involved in the issue even when no proposal was currently under discussion
among policymakers. These organizers also hoped that formal commitment to a
single-project coalition would enforce discipline in the supportive players. Leadership under the informal alliances that preceded both of the new coalitions was
seen as powerless to hold partners together.

Agreeing with the political problem of weak alliances and actually yielding
on distinctive concerns with the issue at hand have proved to be mutually exclu-
sive decision for those who support the principles of both coalitions. Spearheaded
by Linwood Tipton, vice-president of both the Milk Industry Foundation and the
International Association of Ice Cream Manufacturers, the Sugar Users Group
was able to organize only around the sugar program in preparation for the 1985

farm bill. The interorganizational and policy process logistics of dealing with issues of both dairy and sugar proved impossible. The elimination of the more costly sugar program emerged as the only issue against which commodity user industries would commit themselves to work jointly. The dairy program, on the other hand, was supported by many milk users inside the Milk Industry Foundation because these members operated as farmer-owned cooperatives.

When the National Save the Family Farm Coalition (NSFFC) was formed, there seemed to be more potential for a comprehensive alliance of farm crisis groups. Unlike the Sugar Users Group and its partnership of organizations with clearly defined policy objectives of their own, the widely scattered farm crisis and other farm support groups remained broadly reformist in their policy positions. Working with Jim Hightower, several crisis committees put together a proposed Farm Policy Reform Act by incorporating as many farmer complaints as could be identified during the national organizing of 1983. That alternative farm bill was a catchall document in which several proposals were included rather than a series of restrictive tradeoffs in which only a few policy goals were articulated. Despite the optimism of National Save the Family Farm organizers, the alliance was nearly stillborn in 1986 and continuously met a series of bitter policy disagreements. Several representatives of the American Agriculture Movement, an important coalition link to Washington politics, still resented the grassroots ascendency of the farm crisis committees, and AAM as an organization remained firmly behind parity through direct government support of commodities as the critical basis of policy reform.

The many state and regional groups that joined the coalition, as well as others that refused to join, were not easily satisfied by a catchall document. Nonfarmer activists within the affiliate groups made any agreement within the new organization even more difficult because of policy goals that were often primarily antimilitary, social reformist, globalist, or opposed to a capitalist economy. Many of these nonfarm representatives were irritated with proposals to increase farm income through higher food prices, the cost of which would fall to the consumer, a concept attractive to affiliated farm activists who were trying to position their demands within the context of federal budget deficits. Since leadership of the many organizations that made up the coalition was divided among those reflecting these diverse views, the differences dramatically affected the setting of a national political agenda for the NSFFC. Throughout 1986, partner groups entered and left the coalition, never agreeing on any more than the need for federal action on debt restructuring and credit reform. Coalition partners finally endorsed what amounted to the resubmission of the Farm Policy Reform Bill by Senator Tom Harkin (D-Iowa) only because that bill had once gained backing and seemed the only available vehicle for bringing unity to the NSFFC. Again only the most common policy denominator held a shaky formal coalition together even during its formulative stage.

INFORMAL INFORMATIONAL COALITIONS

Given the history of formal coalitions in agriculture and the reluctance of group representatives to give in on what they see as major policy goals, it should not be surprising to find informal coalition arrangements in which participants regularly interact with one another but avoid sanctioned cooperation on questions of public policy. Such informational coalitions are analogous in function to private-sector information industries that have evolved to provide highly specialized, business-specific data.[11] Several lobbyist respondents claim that one such organization, the Food Group, is the most valuable coalition to which they have ever belonged. Founded in 1949 and still active, the Food Group elects officers, sponsors luncheon programs on issues of agriculture, invites public officials to attend, and provides Washington representatives an ideal setting in which to make contacts and sound out both potential friends and foes. The organization could be thought of as more akin to a social fraternity than to a multiple-interest coalition, since no mutual agreements are made on public policies in any explicit sense. Some of the participants believe that the Food Group is politically irrelevant for this reason. Others disagree, arguing that lobbying must be essentially a social activity and that the Food Group provides the starting point for most interinterest relationships.

Proponents of arrangements such as those of the Food Group see coalition building primarily as an opportunity to share information about policy events and circumstances and to develop a consensus of opinion about what should be done. These representatives recognize the limitations of formal coalitions. They know that other forums can be found for sharing information, so, with only a minimal commitment to organization, they foster the Food Group as a neutral meeting ground for those who identify their policy problems as agricultural. Such individuals also start and attend more selective breakfast and sports meetings to encourage patterns of interpersonal contacts that increase political understanding, if not necessarily overt collective action, around such facets of agriculture as international trade. In an alternative format, they provide support to print a newsletter on food and fiber lobbying to call attention to who is involved in what.

Coalition activities such as these are most attractive as low-cost alliances that either require minimal maintenance or that can be organized on an ad hoc basis, and they offer more than basic assistance for those lobbyists who want to know people, what they are thinking, and what policy events are forthcoming. Issues are discussed in enough detail that many participants learn what issues and proposals can be lobbied effectively, what should be left to others, and what should be postponed. In essence, some degree of constraining agricultural policy education occurs as lobbyists talk together. These meetings afford like-minded partner groups a forum for presenting their own problems and a chance to put policymaker guests on the spot. As one Food Group loyalist noted, "While this is an innocuous coalition, more than one government official has learned from its members here that his plans will not fly in Washington."

Though information alliances such as the Food Group can be important for communications among private interest representatives and even have some subtle impact on public policymaking, most lobbyists want some middle ground as an organizational format between it and the formal coalition. "Most importantly," said one lobbyist, "we just need to meet and work out whatever differences we can in our [policy] proposals. It just makes it easier to get something done. Usually it is the only way."

These representatives expressed the opinion that "the greater the consensus, the easier to accomplish something." According to such respondents, coalitions have two purposes. First, representatives hope to negotiate a common position on a detailed proposal that partner groups can support as they work it through the policy process. Second, they want to develop the strategy needed to see the proposal through. The intention of this internal cooperation and the image of unity it creates, as identified earlier by Ann N. Costain and W. Douglas Costain, is to ease the attainment of policy goals by convincing government officials that affected interests are not in conflict with the proposal at hand.[12] In order to create this impression of sweeping support or opposition, informal policy coalitions may come together at any time when organized momentum is judged most effective in securing joint goals. Although such alliances can shepherd a bill through the entire policy process, these loosely maintained coalitions are most apt to focus on just a few stages of decision making: the election or appointment of a policy spokesperson, the development of the proposal, legislative and White House deliberations, program implementation, and even the administrative evaluation of program operation.[13]

Several points made by respondents are crucial to an understanding of the limited role the participants see for informal policy coalitions. Ideally, partner groups aim to develop a proposal that legislators and administrators will ratify.[14] Occasionally, because policymakers want to avoid an issue or cannot decide on an approach within their own institutions, simple ratification of a coalition agreement does occur. In practice, ratification seldom results, because of the perceived independence of legislators, the policy expertise of administrators, or the vast amount of additional information these policymakers also process. Such factors enhance the importance of multiple-interest coalitions; otherwise, inattentive policymakers may feel more compelled to listen to a battery of organizations emphasizing the same position. Still, respondents report, policymakers rarely accept predetermined packages just because several interests agree on the content.

Informal coalitions are both proactive and reactive. That is, they develop policy positions because of the partners' perception of the need for new policies but, more frequently, as alternative responses to proposals that someone else already has placed on the policy agenda. Potential coalition partners less frequently are attracted to new issues, since proposals of this kind have not been around long enough to have gained a following or for circumstances to have developed that convince likely supporters that they can be won.

Interest representatives who are most strongly inclined to endorsing the importance of informal policy coalitions emphasize the informational benefits of common alliances directed toward rapidly approaching policy ends. To these respondents the need for policy suggests that a pending organizational problem demands an immediate solution. Information is gathered only as a means toward a solution. "There are too many people who know everything but do nothing," concluded one food lobbyist. "The meeting of involved groups should be a forum for discussing only relevant information about pending action." This attitude often leads many coalition players to be quite skeptical of the ongoing formation of informal coalitions. Skeptics charge that too many lobbyists become active on a proposal far too late and use coalitions in some quick and dirty fashion "to find out right now everything they needed to understand weeks ago on some complex issue."

Respondents less supportive of informal coalitions often become partners primarily to learn about what goes on with other interests and continue to plan their strategies accordingly. They value and encourage policy-directed coalitions over the Food Group format because information obtained there is more policy-specific and more useful in dealing with pending legislation, but there is also a strong compulsion to join such coalitions in order to remain abreast of rapidly changing events and to make certain that others understand members' positions. For these reasons a strong majority of coalition partners are rather passive. Most passive partners, moreover, feel they are inclined to be less involved than others in most coalitions their organizations join. Given different issues and political conditions, however, representatives who are passive partners in one coalition infrequently may feel compelled to exert strong influence over another, but those who see fewer benefits in coalitions generally are more reluctant to assume leadership role even on critical issues.

Coalition partners, both active and passive, do not agree that informal alliances lead only to minimal coordination and the joint assumption of only a public posture.[15] Successful coalitions, in terms of both organization and policy, emerge only when partners take them seriously and recognize the need for some resolution of the problem that players can support even after the coalition meeting ends. Those who undermine alliance agreements after promising support or neutrality are likely to feel penalized in the future. "People just won't work with you, no lie!" was a typical remark, "You can't afford that in this town." While different partners carry different organizational work loads, they do subscribe to a set of norms about maintaining the agreed-upon coalition position as long as it holds together in policy circles and within the alliance.

The informal coalition gains favor over the formal not because policy goals are different, but because the demands of supporting informal collective action are far lower. Except for mutual honesty, participants do not expect as much from one another under informal conditions. Such coalitions can be organized quickly. Commitments to maintaining the alliance are minimal, dependent only on resolution of a particular issue. Participants come and go and may change their positions at any time. Leadership usually tends to be based on who initiated the coalition. Membership is most frequently by invitation but is generally open to whoever wants to attend. Resources are shared through voluntary contributions. Lobbying assignments are divided among the participants, but monitoring of who does what is minimal. Discussions about how to define the similarity of intergroup interests and negotiated trade-offs on proposed provisions, as well as peer disapproval, are the only real incentives and deterrents to cooperation.

From an organizational perspective, this arrangement is weak, but advocates point out that formal coalitions have functioned little better. In many ways, formality has been less effective because it intimidates potential partners and makes them especially conscious of their own policy turf throughout their involvement with the coalition. Formal coalitions of groups interested in agriculture may well have succeeded in creating potentially enduring umbrella structures housing like-minded but unique and diverse interests in the NCCO and the NFC. However, in a short time, partners vanished, the leadership weakened, and there were few responsive participants. Informal coalitions, on the contrary, keep reappearing, often as the same issue arises in the form of another policy proposal or political problem.

Some long-time lobbyists and policymakers offer additional support for informal coalitions over formal structures. They claim, and historical events seem to bear them out, that formal coalitions initially were unwarranted imitators of what had worked so well with less organizing effort. Informal farm coalitions have been around since the Farm Bureau, the National Grange, and the National Farmers Union first worked together in the early 1920s. Agribusiness leaders first engaged in coalition building when the American Council of Agriculture sought to organize industry and local farm organizations in support of financially destitute Depression Era farmers.[16]

In agriculture, at least, the formalization of interest group coalitions has been little more than a reaction—and many critical lobbyists would claim an overreaction—to the perceived success of not just other influential interests but also of their informal cooperation. When farm groups founded the National Conference of Commodity Organizations, their operatives felt that a tightly knit and well-coordinated set of agribusiness and conservative farm interests had successfully fronted the Eisenhower administration's assault on protective farm policy.[17] Soil bank programs were attacked with vigor by fertilizer, seed, and equipment dealers. Price supports had been and continued to be the coordinated target of these agricultural input groups, the American Meat Institute, the Farm Bureau, National Cattlemen, and several smaller livestock interests. The success of these informal coalitions in getting the attention of agricultural policymakers was so marked that these same interest groups, together with food retailers, later felt confident enough to challenge proposed national economic policies while claiming to be the united voice of agriculture. The NCCO with its even tighter and more coordinated organizational structure was intended to give competing farm interests an advantage by organizing better than did their conservative opponents.

When the National Farm Coalition formally reorganized, this too was in response to representatives of agribusiness interests who, under the Nixon administration, were jointly planning farm bills for the early 1970s that many hoped would deregulate agricultural production in favor of the free market. The National Resources Council was the same organizational reaction to what was seen as better-accepted, better-funded, and better-staffed opponents in industry and agriculture. In the case of both the NFC and the NRC, formalizing the coalition was little more than an expression of hope that policy victories could be gained by outorganizing the opposition. That hope, as veteran lobbyists now see it, was misplaced only because there are limits on how far and how long any private interest can sublimate the goals it derives from its principal services and long-term policy preferences to those of organizational outsiders. The game is played out when individual members quit and declare their private victory or accept the loss and move on. Informal coalitions are well structured to accept this ebb and flow as a natural part of a constantly changing policy process.

THE GROWTH OF
MULTIPLE-INTEREST COALITIONS

Despite both the limitations of multiple-interest coalitions and the frustrations of keeping allies together, private representatives believe coalition activity to be on the increase. Moreover, they see the same three reasons responsible for this increase: more centers of policymaking, greater numbers of organized interests, and a more complex array of policy issues. It appears that coalition building among organized interests, as an evolving tactical strategy of more frequent alliances, is primarily a response to new policy conditions rather than any refinement in the ability of groups and firms to work together more effectively than in the 1950s or 1960s.

MORE CENTERS OF POLICYMAKING

As noted in Chapter 3, congressional control and the administration of agriculture are now more fragmented among multiple centers of policymaking than ever before. More committees and agencies have gained policy responsibilities and, thus, influence over the content of public programs. Legislators also tend to be swayed more by opinions expressed from voters back home, such as realtors or conservation clubs, and they use their positions to serve these constituents. There are more issues in which the White House may intervene and, through increased litigation, more opportunities for the judiciary to exercise control. The result is continued organizational stress in which interests and their representatives are spread exceptionally thin in allocating time, appropriating resources, and developing knowledge of the personal dynamics of policymaking. For many interests, even dealing with the Department of Agriculture or with a congressional agricultural committee may be a new experience.

Interests have responded to this situation by forming coalitions in much the same way as they hired consultants. Representatives look for those employed by other groups or firms who, for instance, know legislators whom they do not, who have access to the White House, or who may have developed expertise in the use of litigation as a lobbying tactic. Coalitions of agribusiness chemical firms, faced with fungicide and insecticide legislation and an individual lack of familiarity with many of the involved members of Congress, first attempted to handle the FIFRA problem by having each firm cultivate the support of legislators with plants in their districts. When the presence of the chemical EDB in manufactured foods became a political issue, food companies pursued the same strategy to use their collective contacts and lobbying tactics to work on the issue as it moved through USDA, the Environmental Protection Agency, congressional investigators, and even state governments. Few coalitions fail to divide up lobbying assignments in this way.

Recognizing the importance of other centers of power, policymaking institutions add to the prevalence of interest-based coalitions. State Department agencies, as well as the Office of Trade Relations, attempt to build coalitions of agricultural group supporters on such issues as the Caribbean Basin Initiative and the international General Agreement on Tariffs and Trade (GATT). After studying the sentiment in Congress for continuing high levels of income enhancement programs in the 1985 farm bill, the USDA adopted the same strategy by organizing the Coalition for a Market Oriented Farm Bill. To some extent, members of the commodity subcommittees within Congress do the same thing on each farm bill when they attempt to structure commodity programs on the basis of a consensus within each commodity's industry. As much as possible, legislators hope to force an agreement between sugar interests or soybean interests so that growers, processors, and refiners for each commodity can all accept the program before final committee deliberation. Such a strategy makes it less likely that opposition from other quarters will appear credible and disrupt the bill's chances of passing.

More Organized Interests

The number of interests representing agricultural issues has made coalition politics more important, since many organizations may appear to have some concern with any one issue. In practice, though the predominant issue orientation of most interests is narrower than is immediately apparent to most outside observers, any policy conflicts that exist among these organizations further complicate the policy process. Competing positions must be sorted out somewhere by someone, and public policymakers can hardly be counted on to facilitate most intergroup coalition building since these officials seldom feel the necessity of passing most pieces of legislation. Rather, since many policy decisions can be avoided, organizations just build their own alliances apart from the scrutiny of government officials. The more organized interests are active, the more sorting through of policy positions becomes necessary, and thus the more coalitions are formed. This phenomenon of multiple interests involved with a single proposal in which coalition participants are likely to be affected differentially has been responsible for the creation of such alliances as the Ag-Energy Users Coalition, the Export Processing Coalition, and the Pesticide Users Coalition in the 1980s.

The manner in which interests are organized according to their type of policy involvement also contributes to the increase in coalition arrangements. Lobbyists identify two organizational types—multipurpose and single-project groups—that encourage coalitions for reasons other than just the need to negotiate a compromise on a policy decision that will have a major impact on members or patrons. Multipurpose organizations, in which the number of policy items attended to usually grows faster than does staff size, both lead and join many coalitions in order to gain credibility in several issue areas. Groups such as the Farm Bureau, National Cattlemen, and Food Marketing Institute are well known as the "big coalition players in agriculture." The NFU and the NFO, with even fewer staff resources, spend little lobbying time on anything other than multiple-interest coalition activity in order to be involved in, and have some influence on, as many pending policy matters as possible. For example, in preparing for the 1985 farm bill, the Farm Bureau and the FMI, along with Public Voice, organized the National Food Policy Conference in order to develop some integrated leadership in a movement to reduce sugar and dairy prices. The three-group coalition was intended to be of symbolic importance in demonstrating the mutually compatible goals of farmers, grocers, and consumers as the "big three generic food interests."

Sector or industry integration is another coalition goal of many multipurpose interests. The Cattlemen continually attempt to legitimate their self-proclaimed role as private interest spokesperson for the entire cattle industry by involving the NCA with problems that come up with smaller feeder or cow/calf organizations with narrower or regional policy goals. Because of the similarities in the problems of other red meat industries, such as hogs and sheep, the Cattlemen encourage intergroup cooperation among organizations representing these animal products as well. The general farm organizations and National Council of Farmer Cooperatives attempt to use such integrative coalitions as their major sources of influence. They see themselves as the only farm groups that represent a diversity of producers and are therefore able to bring commodity interests into discussions of common problems within a shared organizational setting.

Many of the more intensive interests and policy concerns find coalition building to be of far less organizational importance than do the multipurpose groups. Single-issue groups (for instance, cotton, rice, tobacco, and peanuts) are more involved with exchanges in which trade-offs in support win common allies among policymakers. Groups representing these commodities, as complicated as their production problems may be, seldom find any advantages in forging alliances over issues that do not affect the immediate economic conditions of their members, as does the Farm Bureau. These single-issue interests worry little about being expected to serve as private spokespersons for every potential problem their members face. Consequently, coalitions are generally less burdensome and fewer in number for these single-issue interests. As a legislative staffer replied, "Wheat and corn have no problems supporting one another with great regularity because they are not in conflict or competition and their problems are similar."

But not all groups can be so selective about engaging in only the most opportune coalitions. Since they are lacking in recognition, political access, proven lobbyists, PAC funds, and so forth, many of the single-project interests, almost always recently activated and not well established, have no tactical alternatives to coalition building. As a result, these interests attempt to bring their proposals to the political forefront by means of expanding support from group to group and eventually gaining more widespread attention from the media and from policymakers. The mutual assistance that ex/al interests continuously provide one another as each specializes in a current project within a broader issue area, produces a constant flurry of coalition building as relatively less powerful interests attempt to overcome their political disadvantages.[18]

While most of this coalition building is between groups with similar issue concerns, as with environmentalists or hunger activists, coalition activity often expands to encompass more diverse interests. FIFRA and the conservation provisions of the 1985 farm bill represent two typical, independent coalition arrangements involving single-project organizations. The groups that initially worked together with plans to oppose long-standing agricultural positions included, on the respective issues, environmentalists and more traditional conservationists. As both policy problems gained greater attention, and the odds favoring ex/al positions improved, the two coalitions of attentive supporters grew to include both environmentalists and conservationists. In a similar fashion, as the issue of required protective clothing for field hands who make direct contact with agricultural chemicals gained media attention, environmental and farmworker interests forged a unique coalition as their leaders saw a probable policy victory.

Under even more unusual circumstances, a single-project group can help build a more encompassing coalition of diverse interests. The Child Nutrition Forum, once fostered by the Food Research and Action Center's attention to the problem of malnourished children, has grown to include many agribusinesses, producer interests such as dairy and eggs, and representatives of school lunch and related organizations. The success of the forum in keeping the issue alive and the participants together results from the fact that all the partners clearly win as the coalition attains its policy objectives: commodities are used, surpluses of produce consumed, manufactured items purchased, school lunch programs funded, and nutritional goals established.

MORE COMPLEX AGRICULTURAL ISSUES

The third reason for the increasing number of multiple-interest coalitions results from the changing nature of agriculture and, as a consequence, agricultural policy. Agriculture and agribusiness could be summarized in the 1980s as much larger in scale and more highly concentrated than ever before, with the largest farmers responsible for both 90 percent of the production and most of the crop surpluses. Moreover, as an entire sector, agriculture continued to lose many of its most productive but financially troubled individuals and firms. Agricultural policy, designed to maintain farm incomes and provide future production capacity, was not successful in achieving either objective. Yet it was operating at the highest program costs ever and providing windfall payments to many of the least needy producers. The entire sector had become dependent on exports in a period in which major crop surpluses were an international condition and U.S. exports were declining in the face of foreign competition and world financial problems. Ironically, for the first time ever, the U.S. began to import more food—in terms of dollar value—than it exported. This led to a further erosion in the already deficient U.S. balance of trade. Even with these conditions, agricultural policy was troubled by problems of malnutrition, inadequate conservation practices, and such declining resources as water.[19]

The impact of these conditions produced at least six problems for interest groups: keeping informed about changes, understanding the implications of change, adjusting previous policy positions where necessary, concentrating attention on new policy approaches, dealing with policy concerns that were somewhat broader in scope than many previous ones, and thinking about the relationships between what were once thought of as rather distinct policy areas such as domestic price supports and exports or agricultural research and trade policies. By 1985 it had become clear to respondents that American agricultural policy was no longer able to determine conditions in a world in which many countries' agricultural systems were interdependent and in a country in which competing interests shared the political attention of uncertain policymakers.

The complexity of these problems and the array of changing agricultural issues, each more closely related than has been evident in the past, leads to many new coalitions of multiple interests. Established organizations with both multiple- and single-issue interests need to worry about the financially fading farm sector, commodity overproduction, declining exports, changing patterns of production among importing nations, the impact of technological innovations, and the degree to which adequate production levels can be sustained. Often the ideal solution to one problem worsens the conditions surrounding others, as for example when assistance for new technology increases commodity production in an already glutted market while simultaneously raising concerns about future production failures. One perplexed respondent summarized the current situation, "Every one of the larger [farm] organizations in this town must spend an inordinate amount of time becoming familiar with conflicting agricultural developments and looking, with other groups, for policy answers that we could not have foreseen only five years ago." For this individual and for other lobbyists, a changing world agriculture meant more coalitions as their organizations began to adjust their respective political agendas to problems that are still thought of as primarily someone else's concern but that can no longer be entirely ignored.

In this context multiple-interest coalitions provided three important benefits for partners. First, meetings became sources of information, where facts could be analyzed and rumors sorted out. Independent staffs were too small and busy elsewhere to focus on comprehensive policy analysis. In several instances, coalitions were able to share resources to employ consultants who examined problems of joint concern such as European Economic Community subsidy programs or incentives for Argentinian crop production.

Second, coalitions often provided the only way to mobilize enough intelligence about the various aspects of a policy problem to address a proposal currently before Congress or an administrative agency. For example, the Coalition to Reduce Inflated Milk Prices won considerable attention because different coalition members were able to address industrial production costs, noncompetitive U.S. prices, dairy surplus conditions, abuses in surplus distribution programs, and antitrust concerns related to the regulation of milk co-ops. Coalitions of grain-trading firms, value-added food manufacturers and processors, and conservation groups were able to advocate specific proposals during 1985 farm bill debates only because information was shared and jointly analyzed.

Third, coalitions became organizational enterprises when a new awareness of policy conditions arose. In the resolution of FIFRA, both chemical companies and environmental groups yielded and became mutually supportive of one another after looking at the potential consequences of no new bill. In 1986, after cotton- and rice-marketing loans appeared to be generating increased international sales of U.S. commodities, a coalition of organizations, grain traders, and other agribusinesses began to put considerable pressure on the USDA and Congress to extend these programs to wheat and corn. On their own, as 1985 commodity legislation was being drafted, the national wheat and corn associations had rejected the marketing loan concept in favor of smaller changes in ongoing programs. However, after operating for just over six months under the 1985 legislation, joint analyses by various trade-dependent interests caused a reversal in the policy positions of each of the partners, including organized wheat and corn growers.

Does one reason for increased coalition building outweigh another? There are important linkages between a changing agriculture, new policy directions, newly active interests, and a more involved set of institutional actors within government. Some respondents see new groups identifying new issues and nontraditional agricultural policymakers as responsible for new items on the political agenda. The immediate effect of increasing coalition politics is that established interests respond to the pending problem of working with the new group or the recently involved public official. However, most interest representatives agree that without a changing set of agricultural problems, institutional changes would matter little. "Despite what it may seem," one lobbyist noted, "we are not engaged in endless gamesmanship here. New players and new rules are frequent, but they reflect changing conditions that few of us were prepared to meet on our own." Increased coalition building among agricultural interests apparently results from both substantive and procedural policy changes that mandate tactical shifts to cope with the political implications of a transitional American agriculture.

COALITION BUILDING

The irony of coalitions, at least from the perspective of the participants, is that on more complex policy problems few interests can live without them. These interactive alliances are simultaneously restrictive and useful in that they provide a means for combining resources for lobbying. In many instances, coalitions provide the only way to remain involved in the resolution of a policy issue. An intensively focused agricultural interest such as the Environmental Policy Institute would prefer the opportunity to capture the attention of Congress and federal agencies on its own in developing legislation and regulations on biogenetic seed production, and a multipurpose organization such as the Farm Bureau retains the lobbying infrastructure for a frontal assault on Congress on behalf of an alternatively proposed farm bill or trade policy.

The political circumstances are the same for both groups, however. Neither has the credibility among legislators, as a critical private interest with any semblance of monopoly control over relevant information, to gain what it wants. Both the representational institutions and the range of related agricultural issues are too complex and too narrowly specialized. A policymaker or another private interest, either because of individual desires to influence an issue or simply because of a different goal for public policy decisions labels the AFBF "the voice of the large rich farmer" or the EPI "the prophet of gloom and doom." As one lobbyist skilled in intergroup rivalries explained, "It's our way of disenfranchising anyone's claim to absolute control over an issue, and it's easy because nobody likes a know-it-all son of a bitch."

Because of the complexities of agricultural policy and its policy process, the fragmentation of interests into substantive and procedural policy niches provides the opportunity to engage in issue leadership but not the chance to control or dictate policy decisions. No organized interest, on its own, can structure policymaking situations in such a complex environment to get its way consistently. "Everyone," said one lobbyist sadly, "needs a little help from their friends even if you have to make new friends to get help. The time has passed when I can get into any office I want. Now I need other [interest groups] to cover those bases."

Making friends or building coalitions remains a difficult enterprise despite the perceived importance of this lobbying strategy. The independence of private interests, distinct policy goals, and separate organizational structures certainly interfere. Personality and egos also get in the way.

How are coalitions formed then? No organization simply comes together. Incentive theory is important to understanding the dynamics of coalition building; potential partners need to be attracted to an alliance that imposes costs upon both the participating organization and its representatives. Someone must structure the exchange of information and the opportunities to coalesce around attainable policy objectives. Entrepreneurs, as Robert H. Salisbury and James Q. Wilson identified them for interest groups, are critical to developing these relationships in that there are, as another lobbyist stated, "good salesmen who get us together."[20]

Coalitions offer a somewhat different set of participant incentives than do individual interests. Groups may offer selectively received insurance benefits to gain and attract members. Coalitions, in contrast, offer group representatives lobbying assistance that makes their workload easier because information is gathered more quickly and specific contacts are made for them. Respondents also view many of the material rewards of coalitions as unique, in that participants gain certain tactical advantages that are otherwise unavailable. Several benefits were noted: organizations can present their demands as representative of a variety of organizations rather than as a self-serving interest; with multiple-interest support, even new and untested proposals can be made to appear as if they are part of the mainstream of political ideas; representatives can often participate within a coalition to become involved in a policy decision with anonymity from members and certain policymakers who would be offended by open support of a proposal; and coalitions of diverse interests, such as farmers and consumers, can work to cultivate broader support within government by simultaneously promoting and identifying their common concerns in different policymaking centers as, for example, farm legislation or consumer legislation.

Coalitions also provide other incentives, which, while intangible, are similar in the way they help build both group alliances and individual groups. Solidary rewards, for instance, are especially important because many participants see considerable glory in working with well-known partners. Sometimes these relationships are seen as opportunities to gain new jobs or at least upward mobility as an insider in a profession relatively difficult to penetrate.

Other lobbyists gain considerable satisfaction in resolving differences with professionals who understand their political problems. Unsympathetic policymakers, attuned to the same issue concerns, often treat lobbyists as part of the problem rather than as the problem solvers they might prefer to be considered. Finally, coalitions offer participants a variety of rewards, including the opportunities to emerge as leaders on an issue of personal concern or simply make others in the alliance listen to a particular position that would otherwise find no audience at all. In short, while many lobbyists may find coalition activity distasteful, they may gain a variety of personal rewards that are not necessarily of much policy consequence but that do get and keep partners allied.

Not surprisingly, coalition builders have become instrumental participants rather than mere common partners to the alliance. Interorganizational brokers have emerged with entrepreneurial skills to facilitate and sometimes even specialize in arranging coalitions. Two patterns of brokerage are evident in agriculture, one dominated by organizations of specific types and the other highly personalized.

Organizational brokers are found among those interests previously identified as the "big coalition players," the Farm Bureau and several other multipurpose organizations. These are groups and sometimes firms that provide many selective services and are active on numerous issues. Their policy niche owes itself to the breadth of organizational concern. As such, these organizations are perfectly positioned as umbrella interests under whose leadership a number of other related interests can come together. The Farm Bureau can accommodate all of the commodity groups, many of whose members are Farm Bureau members as well. The Food Marketing Institute can accommodate large retailers, small grocers, national supermarket chains, and more specialized outlets. The National Council of Farmer Cooperatives is linked closely, through the co-ops, to such diverse producer/middlemen as those in dairy, cotton, rice, and citrus. One lobbyist respondent called these multipurpose organizations "master of absolutely no policy positions, watchdogs of quite a few." In order to exercise any degree of continued political influence and policy prominence, these organizations broker coalitions.

While much of their organizational brokering goes on in terms of relationships with groups having an overlapping membership, brokering also becomes a responsibility that brings these multipurpose groups together, even with competing interests. For example, the National Council of Farmer Cooperatives can promote the idea of marketing loans as a price support mechanism and then leave it to co-op-dominated cotton and rice interests to develop operational programs with the 1985 farm legislation. In contrast, however, brokering organizations such as the AFBF and FMI are well organized to address jointly several major provisions of that 1985 legislation and even work with a narrow consumer interest such as Public Voice. When other groups such as the National Cattlemen's Association expand their issue involvement into tax policy and nutrition, they have little choice but to foster coalition arrangements. To do otherwise exposes their lack of familiarity with both the substance and policy dynamics of the issue, since the NCA and other new players can hardly claim instant expertise.

Individuals who work for groups playing a brokerage role actually do much of the coalition building. Some of these representatives are effective at negotiations while others are not. As a consequence, depending on the issue and the broker, some coalitions may be especially strong or weak in mobilizing behind the matter for which the participants have allied themselves.

Personalized coalition brokerage leaves far less to chance. Several agricultural interests, such as the FMI and the Agriculture Council of America, hire some representatives on the basis of their brokerage skills and contacts. So too does a less active coalition player, the Grocery Manufacturers of America. As a result, the work of numerous lobbyists in Washington revolves around the cultivation and maintenance of these often necessary alliances. The employers' attitude can best be summed up: "If we're going to need coalitions of interests, at least we'll have them set up by our own people with our interest in mind."

Much of a consultant's time, both for lobbyists and technicians, goes into personalized coalition brokering.[21] Multiple-client lobbyists who consider themselves issue managers see coalition building, either on major pieces of legislation such as the farm credit bill or on single legislative provisions of the farm bill such as a wheat reserve, as one of their major responsibilities. Many interests lack brokerage skills, experience in developing cooperative relationships with other organizations, and knowledge of political affairs beyond their narrow confines, so external brokering, based upon situational needs, is in high demand by many groups, especially more policy-intensive ones that, under the nonexchange coalitions noted earlier, are breaking new policy ground and can afford the assistance.

The result is a fairly heavy concentration of agricultural policy actors and organizations that regularly broker multiple-interest coalitions as important lobbying responsibilities. Coalition building, according to respondents, is dependent on their contribution. There is little evidence that these brokers are exceptionally successful in winning, however. There are too many variables in the political arena for coalition tactics to be a panacea. More importantly, since many coalitions change, many alliances still win on a proposal but gain few of the benefits initially sought by the original partners. Moreover, there is a consensus that most of the multipurpose broker organizations are less influential than those interests they help bring together. For example, the reputation of the single-issue commodity groups for getting the policy results they want is far greater than that of the Farm Bureau. Likewise, Safeway and Archer-Daniels-Midland or Stanley are seen as more likely to succeed on their individual goals than are the FMI or the Corn Refiners. The only conclusion that most lobbyists can agree on is fairly simple: coalitions are important tactically in channeling attention to major policy issues, but it is still easier for private interests to win smaller and less encompassing victories. In the words of one agricultural policymaker, "It's just easier to set commodity policy than it is farm policy."

10

Issues, Interests, and Public Officials

The notion that organized interests can be understood in terms of their occupancy of certain policy niches lends an important degree of order to the loosely connected system of private-interest representation in American agricultural policymaking. On an individual organizational basis, it also reveals an interplay between what interests do for their members or patrons in both service and policy. For example, an organization that emerges to perform services for food retailers will emphasize marketing and sales problems in its lobbying. Likewise, a firm built by its executives on exceptionally speculative risk taking will look for governmental action to buffer against financial losses rather than serve as a private spokesperson for market-oriented policy. These concerns, and little else, will bring nonfarm interests into policy debates over commodity programs, and, even then, their involvement will be self-restricted to these points of emphasis rather than broadly reformist. For example, far too much has been made of the Farm Bureau's insurance program, a service designed to attract members to the AFBF.[1] Though these links have been important, the Farm Bureau has had to create a separate organizational structure to manage and, most importantly, represent insurance programs politically. Farm lobbyists, with a farmer constituency, were never much concerned with the problems of the insurance industry. Such dynamics also explain why coalitions of private interests are suspect, short-lived, and, in words favored by several respondents, "always shaky." Departures from a policy niche are of questionable value and potentially of high cost to an organization's policy identity.

These policy niches must be explored more carefully, however, especially in terms of the substantive policy impact and political influence of those who occupy them. The temptation to label them policy domains, rather than niches, exists because it implies some jurisdictional control over specific programs and proposals.[2] But, with so many other influence agents at work politically, no one can claim that organized interests exercise any absolute control over the resolution of issues. The idea of occupying a niche at least implies that there are distinct issues that certain interests and few others will attempt to influence routinely, a factor that throughout this study has been shown to account for the reputation of an interest as a private player in national politics. But are organized interests successful in their heartland issues? And how extensive is their involvement in matters that seem to be their obvious concern?

To this point, the first emphasis of earlier chapters has been on determining policy position—that is, who represents what in a changing agriculture? The chapters have also examined tactics for exercising influence as they are affected by the organization and structure of the agricultural lobby and its coalitions of interests. Of the seven questions outlined in Chapter 1, all but those two concerned with policy impact have been addressed. They are the topics of the next two chapters, this one on ongoing policymaking involvement and Chapter 11 on the politics of omnibus farm bill legislation.

This chapter seeks to measure the involvement of private interests to draw some conclusions about their importance in determining public policy. Direct measures of influence have proven to be so elusive that researchers must look for secondary evidence, including perceptions of success. To satisfy readers more specifically concerned about agriculture than academic questions of political power, the chapter also advances the investigation of who represents what positions and why. Relative to an interest's policy niche, or the locus of its political involvement, two questions arise. To what extent do actual policy demands reflect the rhetoric that characterizes interest group politics? How frequently are policy decisions that specifically relate to an organized interest's political purpose made without its involvement? Framed together, these questions ask: When does the public policy involvement of organized interests have a policy impact rather than simply reflect some organizational need to posture on an issue?

PRIVATE INTERESTS AND PUBLIC POSTURING

Interests take the policy positions they do for many reasons. Multipurpose agricultural organizations, especially the general farm groups, have extensive checklists of positions on a wide variety of political issues. Each organizational book, as the compiled statements are often called, results from some agenda-setting process involving the members or corporate officials. While the membership of the American Farm Bureau Federation sets its agenda through a series of county, state, and national meetings, a firm like Cargill responds to the concerns over public policy activity that may have caught the attention of numerous division executives. Organizational harmony demands that no collectively sanctioned agenda item be unceremoniously buried, even though it may have little to do with farming or trading gains. While the Farm Bureau circulates its multi-paged policy statement, some Cargill executives wait for the opportunity to press their case through letters, personal legislative contacts, and prodding of the public affairs staff. In the case of these and other private interests, some form of executive committee composed of more politically informed officials—or at least more authoritative ones—decides that some member wishes will gain priority status while others are inappropriate to pursue on an active basis.

Life is little different for the representatives of other types of organizations, except for the need to respond to fewer issues. While the protest group's supporters agree on fewer positions, the single-issue and single-project groups have more restricted goals that decision-making bodies within the organization have agreed to follow. As a consequence, these organizations seldom advertise extensive lists of public policy concerns. That does not keep various members and patrons of even these interests from demanding greater responsiveness on their own personal priorities from those who work for the organization in Washington.[3] To appease such claimants, organizational staff, as policy gatekeepers, make a great many proclamations and offer several critiques of government that may be entirely unrelated to the policy positions they assiduously research, lobby for, and actively represent to policymakers.

Organizational representatives also adopt a public posture on many issues to support those in policymaking circles who have assisted them. As noted in the previous chapter, many coalition partners play this role passively. Groups and firms frequently testify on behalf of programs, publicly endorse proposals, and enlist in a cause such as tax reform even though these matters have no priority status. An organization may want to maintain close political relationships, promote unity among those who must regularly work together, or help keep a favored individual in office. Position taking, rather than serious lobbying or a concerted public relations campaign, is the result.

Relationships with both financial and political supporters are maintained in numerous ways, most of which can be observed quantitatively. Testifying at hearings is frequent, both in Congress and when administrative regulations are being considered. Comments to the media and statements made through in-house publications are used by most interests under these circumstances, since these are highly visible and relatively easy to issue. Calls to friendly legislators and other allies, such as constituent members, are less frequent because they imply stronger support and can be personally misleading. "Don't, don't ever, help a friend by confusing things for another," cautioned one lobbyist. "Let people know your priorities by the way you stand behind the positions you take."

Posturing to maintain supporters, because of its highly public nature, creates difficulty for those concerned with measuring or judging influence. Much of what can be seen is irrelevant to what an organized interest really wants. Moreover, much of what is done—cultivating personal contacts, mobilizing constituents, presenting research findings—to attain its goals can be readily seen. No written records or paper trails are maintained for the benefit of those who may wish to comment on an organization's influence.

But how frequently do interests engage in public posturing? Respondents note that all organizations, and all representatives, do it with some frequency. While time-consuming, these demonstrations of support seldom interfere with more urgent tasks. "If we can't do it, we just don't." said a lobbyist. "People recognize that, and, if they don't, we explain our priorities." Nonetheless, all respondents agreed that posturing consumes valuable organizational resources that could be utilized elsewhere.

More importantly, does the prevalence of public posturing confuse the policy process by creating a false image of consensus, competition, or conflict? This creates the most serious problem for researchers, a group sorely lacking in insider information as to what is just a public proclamation and what is a serious policy claim. The problem can be seen in the contradictions of scholarly findings. Using either perceptual data on the importance of interest groups or unsystematic observations about issues involving organized interests, various academic studies are in disagreement. Case studies often show examples of conflict that support the idea of a pluralistic policy process in which competing interests champion their causes.[4] However, other studies suggest that conflict between interests are seldom typical of public policymaking in the final resolution of proposals.[5] These studies are supported by analyses of the congressional hearing process that demonstrate considerable solidarity among most of the organized interests that show up to address an agency's funding and programs.[6]

What are the realities of organized interests' articulating so many political demands? Is conflict, as so often seen through contrasting statements, real competition over program objectives or simply a false impression created by public posturing? Are demonstrations of unity expressions of widespread endorsement of certain programs or merely the acquiescent comments of organizational representatives who are maintaining group or firm relationships?

A simple measure was developed to analyze the extent to which single issues were the subjects of posturing by interest group representatives. First, the number of organizations publicly testifying or commenting through the media or in-house publications was ascertained. Second, group representatives were then asked to describe their involvement with that issue. Finally, a percentage score was calculated to show the number of interests actively lobbying on the issue.

One issue area selected was agricultural research. Specifically, which interests were opposed to current directions in federally funded research programs? This issue was useful to examine, since Don F. Hadwiger had outlined the historical patterns of conflict and because a growing number of comments from other organized interests were being directed toward new research developments.[7] An analysis of the intentions of involved organizations could determine the potential for escalating conflict since industry and university groups as well as the Farm Bureau were known to provide a strong base of support for the research establishment.

Five distinct complaints about agricultural research were noted from policy statements made between 1983 and 1985: the programs cost too much, large-scale producers benefit disproportionally, research promotes inferior and unsafe food products, research produces unsound environmental use of soil and water resources, and the agricultural research agenda is controlled too tightly by government and federally funded scientists who are unresponsive to private interests. These criticisms were voiced by more than just the expected watchdogs from the ex/al environmental lobby. Farm protest groups, church-related rural organizations, ex/al rural organizations, some traditional farm organizations, agribusiness trade associations, and firms developing biotechnology were among the critics.

The complaints were not voiced jointly, and no coalition was organized around agricultural research reform. Farm organizations were concerned with the family farmer. Trade associations were worried about consumer acceptance. Church groups saw the issue as part of their concern with agrarian reform. Altogether, twenty-five respondent organizations were responsible for these diverse complaints, but few shared the same rationale for making their criticisms.

How serious were the perceived opponents? Only five organizations took any action beyond their public statement—including lobbying on related farm bill provisions. The Environmental Policy Institute conducted a large-scale investigative research study. One protest group engaged in a small public relations campaign in the Midwest against biogenetic production increases. One agribusiness interest questioned the findings of university researchers before the Environmental Protection Agency. Two others were developing a strategy for research reform through joint ventures between the public and private sectors. Altogether, on even an issue of broad scope in which several different private-interest agendas could be served, only 20 percent of the articulated opposition to present practices appeared to be real.

None of the other twenty respondents saw agricultural research reform on their actual lobbying agendas, at the time or in the foreseeable future. Only four of these representatives saw the issue as one that their organization would be likely to address again publicly. Why had the matter been addressed initially? Two organizations responded in support of issues raised by a friendly member of Congress, four in response to administrative supporters. Ten respondents believed that the comments made by their organizations were supportive of other private interests. Both a trade association and a farm organizations responded to

TABLE 10.1
ACTIVE INVOLVEMENT OF ORGANIZED INTERESTS
TAKING PUBLIC STANCE ON SELECTED ISSUES*

ISSUE	NUMBER OF ORGANIZATIONS	
	With Stance	Actively Involved
Opposition to agricultural research	25	5 (20%)
Support of ethanol subsidies	23	5 (22%)
Opposition to sugar/dairy programs	30	9 (30%)
Support of stronger FIFRA	16	3 (19%)
Total for all issues	94	22 (23%)

*Includes only organizations represented by respondents

membership pressures. Twelve more representatives, including some voicing support of old allies, thought that agricultural research was simply a good side issue to address since it was consistent with other policies that their organizations addressed actively. As such, the criticisms were useful to include in organizational publications but were illusory as far as understanding an organization's policy purpose and what its representatives wanted to influence.

Three other issues and policy positions were also examined to determine whether such large discrepancies between articulated concerns and actual lobbying were typical. These included support of ethanol subsidies, reduction in federal sugar and dairy price supports, and stronger FIFRA regulations.[8] In each instance, a variety of organizations, ranging from sixteen to thirty, as can be seen in Table 10.1, had testified or otherwise offered public commentary. All had been prominent legislative issues in the mid-1980s and at least some coalition activity mobilized around each position. Nonetheless, only 22 percent of the ethanol supporters, 30 percent of sugar and dairy opponents, and 19 percent of FIFRA supporters lobbied or otherwise became involved actively on behalf of the issue on which their organization previously had taken a public position.

For each of these issues, patterns of active involvement were easier to comprehend. While many producer interests expressed sentiments on behalf of ethanol supporters, only the direct beneficiaries of public programs worked for them. Only heavy commodity users and independent dairy-product-manufacturing interests lobbied against price supports. All active FIFRA opponents came from the ex/al environmental lobby. The issues, however, were less complex than research reform, which directly affected a great diversity of interests.

Public posturing is extensive and explains much commentary that surrounds the policymaking process, especially the private interests without a considerable stake in the political decision at hand. Within the agricultural lobby, the conflicts apparently are not as extensive as they appear to be, nor is support for a proposal as strong or broadly based as it seems, even when backed by a well-publicized coalition.

While most private interests fall back on those few issues and policies that define their public mission or niche within the representational process, there are still decision-making consequences of such posturing. The more it goes on, the more it confuses the process, leading former policymakers such as Don Paarlberg to decry the "babel of voices."[9] Even lobbyist respondents acknowledge that regardless of how hard they try to keep their policy priorities evident, confusion results. Since personal contacts can be made far less frequently than most lobbyists would like, policymakers gain much of their understanding of who wants what from the same official transcripts and media reports as does the public at large.

Misperceptions, especially when they benefit some other organized interest, abound and are often encouraged. Organizations with multipurpose policy goals, that make the most frequent demonstrations of support suffer the greatest negative consequences. They appear to lack specific and clearly defined agendas, often failing to communicate adequately their positions on major issues. Since posturing is commonly done on behalf of a selective set of supporters, the organizations often seem overly ideological or partisan as well. Finally, needless posturing may offend potential allies who neither understand nor need more public criticism.[10] As a consequence, the rhetorical characteristic of interest groups not only complicates the policy process, but it also serves to divide it further and make resolution of existing conflicts more difficult. Thus, while the tendency of organizations to identify and keep to their own policy turf orders the system of interest representation, that tendency contributes far less order to the legislative and administrative dynamics of the policy process. The differences between rhetoric and actual policy demands cannot be perceived easily or adequately by those who are not close to an organization's inner dynamics.

BEING INVOLVED

Observers of organized interests who wish to understand their public-policy influence must be aware of how much political participation is structured by the need to satisfy members and aid political supporters rather than to affect government decisions. While rhetoric and posturing may well assist others in their programmatic goals and be an important policy factor, most articulated demands of interest groups and firms bear little relationship to an organization's own internal priorities. These instances of posturing are not ad hoc reactions to policy situations and events to which organizational representatives suddenly believe they need to respond, even though they are not on a stated agenda. Ad hoc reactions do occur, and they elicit serious responses. For instance, on the defeat of a mandatory production control referendum during 1985 farm bill proceedings, well over three-quarters of the respondent organizations that had publicly disavowed the amendments engaged in serious lobbying as these amendments came up for votes. In addition, many other interests actively lobbied against the same amendments, even though their representatives had taken no earlier stand. Posturing on issues is a far different response in that, as goal-directed behavior, it relates to the maintenance of organizational relationships rather than to the direct pursuit of what an interest wants government to do.

These findings imply that an organization develops a self-identity as a result of what its representatives and staff believe that they, as policy experts, should be doing for either the most adamant demands of those who pay its bills or for those to whom they otherwise provide policy and personal services. That does not mean, however, that these definitions and the attendant emphasis on certain public policies and issues do not change. As noted earlier, representational changes within the agricultural lobby go on continuously as both the political environment and organizational conditions are altered. A transitional agriculture further encourages changing policy responses.

An organization's definition of its policy niche does not mean that its representatives will not lobby actively for legislation or for regulations that they favor on purely personal grounds. Clearly, entrepreneurs and managers have discretion in what they do, given their isolation in Washington and their immersion in political life.[11] Two former farm group presidents provided considerable evidence of that independence as they became national leaders for causes beyond the confines of farm policy. Charles B. Shuman of the Farm Bureau emerged as a frontline spokesperson for conservative economic and hawkish foreign policy decisions, and the National Farmers Organization's Oren Lee Staley became a national proponent of labor-style collective bargaining. In a sense, these leaders transcended their organization's purposes in much the same way as the United Auto Workers' Walter Reuther did in becoming a national champion of progressive social and environmental policy.[12]

But too much can be made of entrepreneurial freedom in terms of an organized interest's actual policy goals and influence. That freedom is hardly unrestricted, existing as it does within the context of specific organizational purposes and identities that others understand to at least some degree. In the first place, Shuman and Staley remained farm leaders rather than economic leaders. They were thought of as farm spokespersons by policymakers and other group representatives, and they gained recognition for their expertise and political followings within that policy setting.[13] Second, these individuals and Reuther, through personal leadership, directed their groups toward multipurpose organizational styles in which staff members focused diffusely on a variety of issues that came to be understood as part of the Farm Bureau or the NFO's policy charge. That is, their groups approached farm policy broadly, eventually allowing a great deal of entrepreneurial latitude for officials. More intensively directed single-issue and single-project agricultural groups have not been noted for their freewheeling independent executives. Third, these farm leaders gained most of their recognition on nonfarm issues by participating in coalitions of business and labor. Except for electoral campaigns, neither the Farm Bureau nor the NFO, as a coalition partner, was a major player or an instrumental strategic component. Like the National Farmers Union, they were primarily vocal supporters, instrumental in attracting a declining rural electorate and useful in giving the appearance of greater national unity on major policy issues. Beyond the contextual dimensions of farm policy, Shuman and Staley gained the most attention for political efforts that won and maintained support for the AFBF and the NFO proposals. In contemporary agricultural policymaking, entrepreneurs have proven far more useful in defining the focus of a policy initiative. As seen earlier, this was the case with protest leaders' mobilizing farm activists and agribusiness staffers who were hired for their farm policy expertise. Both sets of leader/entrepreneurs brought their organizations into commodity decisions, but in neither instance were these actions anything other than a reflection of basic economic problems of the members as understood to be affected collectively by current farm policy.

In their policy pursuits organized interests do not range much beyond their identified niche. The National Cattlemen's Association, for example, does not address nutritional issues in any active way until it gains the lobbying capacity to do so. The National Association of Wheat Growers addresses wheat programs, and the Pork Producers the problems of the hog and pig industry. To see either of these organizations mount a serious lobbying effort on such important but commodity-related issues as tax reform or the Plant Variety Protection Act would be surprising.

If the above parameters seem likely to influence the upper limits of an organization's lobbying effort in terms of the issues its representatives will attempt to influence, what defines the minimal criteria for involvement? Will the National Corn Growers Association be involved on all public policy proposals before government that specifically address corn production? Will all hog proposals involve the Pork Producers? Will multipurpose groups such as the Farm Bureau, with its announced aim of dealing comprehensively with agricultural policy, be nearly omnipresent?

Each of these questions has been opened to interpretation. On the one hand, interests are at least expected to be watchdogs for relevant policy actions. Does this require enough involvement at least to state a position on each proposal as policymakers consider them? The subsystems interpretation of policymaking, with its emphasis on close and continued working relationships among all participants, would imply such extensive policy involvement. Political observers who see the United States as a narrowly organized "interest group society" in which so many groups are organized so effectively to benefit nearly every constituency might well expect the same.[14]

Other political observers would expect interest groups to ignore many proposals that might logically appear to be their business. Policymaking, as it has so often been described, involves many actors who bring many procedural problems and responsibilities to the process. Many proposals may best be left to bureaucrats and legislators, especially in agricultural policy. Bureaucrats are given broad discretion to implement farm policy statutes, and they have extraordinarily high levels of expertise. Congress is noted for its specific attention to farm problems and its limited interest in imposing penalties on those involved in various facets of farming and agribusiness.[15] As a result, interest groups may play—or only may need to play—a far more limited policy role than their harshest critics would suggest.

The extensiveness of private interest participation cannot be predicted by looking at policy influence as an either/or situation, though. Both advocates of the pressure group model of policy dominance and those who see organized interests more as a benign political phenomenon may conceive of interest involvement as relatively pervasive. They disagree only about whether groups fail to lobby because they find the issue mundane or because policymakers act as if their claims are illegitimate.[16] Other literature must be used to explain the degree to which organized interests become actively involved.

Bauer, de Sola Pool, and Dexter provided an important starting point for this analysis in developing their transactional theory of interest group behavior. As was noted in Chapter 3, so as not to be viewed as offending interlopers, they exercised considerable restraint in their involvement.[17] Michael T. Hayes theorized, and later research confirmed, that certain policy types were more the prerogative of organized interests, while other influence agents prevailed elsewhere.[18] For example, interests were involved especially with redistributive issues where someone is likely to lose the benefits of public programs. Rank-and-file legislators tended to avoid advancing redistributive policies and the conflict inherent in them.

Other factors confuse expectations about the lower limits of active involvement by organized interests on issues of obvious concern. Another conclusion of the Wiggins research that suggests extensive involvement is the finding that, on issues in which other influence agents dominate, organized interests consult actively with policy participants prior to final resolution in order to negotiate the best deal possible.[19] On the contrary, Bauer, de Sola Pool, and Dexter find that organized interests often fail to negotiate because they do not understand their self-interest in many proposals that seem to affect them directly. This, of course, leaves them with little to negotiate and limits their involvement.[20] If policy niches are important in explaining why an interest gets involved on an active basis, then competing expectations about when organizations will participate must be explained and resolved.

MEASURING POLICY INVOLVEMENT

The evidence considered throughout this book, based on respondents who discussed their organizations' policy emphasis and the reputations others perceived for these interests, suggests a fairly orderly representational process in which groups and firms construct purposeful political agendas. These agendas constitute only a small part of the concerns articulated in largely public forums. One important factor frequently discussed by respondents is the likelihood of winning on an issue. Organizations are most likely to structure their agendas on the basis of winnable policy demands, and an ability to win seems to be greatly determined by an interest's expertise, recognition on the issue at hand, and acceptance by policymakers.

How accurate are these generalized perceptions of priorities and reputations if one looks at actual involvement on a sample of specific proposals raised over time? What can be confirmed, or what additional information can be determined, about the behavior of agricultural interests if they are examined in the context of the day-to-day legislation and regulation that goes on over time?

Two uncertain conditions that have been discussed in this chapter cloud an understanding of what gets on the agenda that organized interests represent. What is the unlikely possibility of entrepreneurs moving an organization's involvement in distinctly new directions? And to what extent do groups actually address all those issues in which they seem to have an obvious concern?

To test more systematically who became involved in what, a relatively simple research design was constructed and followed. Data were collected on all USDA and food-and-fiber-trade-related executive orders and regulatory activities and on congressional agricultural policies over a five-year period, mid-1980 to mid-1985.[21] Administrative rulings with a specific policy impact, as opposed to rewriting procedures or shifting personnel, were isolated and a sample of just over 30 percent (140 items) was selected at random. A sample of just under 50 percent (40 items) of all farm, food, fiber, and related trade bills passed by Congress was also drawn randomly.[22] The intent was to limit these bills and rulings to mainstream agricultural issues in order to make the data collection manageable. It also allowed for a test to determine to what extent these matters were really the prerogative of the producer portion of the agricultural lobby.

The second step was to ask respondents what decisions their organizations became involved with. These questions were asked from the list of items (see Appendix C) toward the end of the interview after establishing the type and basic concern of the organization being discussed. Respondents were asked about policy decisions that seemed to be of interest and about several that did not. Then they were asked to recall other policies that had attracted the attention of their organization, and these policies were compared with the checklist of sample items. Representatives of reform and multipurpose organizations were asked about far more decisions than were those from the intensively directed single-issue and single-project types. Finally, when an organization was not involved in what seemed to be a relevant decision, the respondent was asked to recall, if possible, where that item originated. Information on whether the group or firm actively lobbied, issued a public statement, or simply watchdogged the proceedings was also collected for each decision in which the interest had been involved.[23]

The data are indeed revealing as supplementary information about the agricultural lobby, especially in terms of understanding policy impact. While the study design did not allow the collection of data on the involvement of all organized interests, it did allow for comparisons of the involvement of major interests and some minor organizations. Thus, it was possible to determine which organizations were most involved in what might best be called "basic" agricultural policy, the degree to which they participated, and the extent to which they perceived other policy participants to be in charge of, or at least exercising leadership over, specific policy decisions.[24]

EXPECTED PLAYERS

Of the 180 legislative and administrative items, only 6 were not of some obvious concern to an organized agricultural interest. Legislation for commercial guayule development and such administrative actions as the granting of most-favored-nation status to China did not reward any specific constituency. On the other hand, legislation such as the Caribbean Basin Recovery Act affected several organizations that had lobbied on various food trade issues of the region. And kiwi producers, only slightly less narrow in their interests than guayule merchants, were organized and had retained Washington representation. With at least 98 percent of the policy decisions of USDA samples and the Congress potentially affecting the activities of an obviously organized interest, agricultural policymaking goes on in an environment in which some interest is always poised to intervene.

To gain an idea of who the players are, Mancur Olson's analysis of organized interests and their relation to public and private goods was modified.[25] As all policies assign benefits to a constituency, sample items could be assigned to one of two categories. Benefits can be targeted selectively to members or patrons of organizations representing a specific facet of farming, industry, agricultural support, or social policy reform. These targeted benefits, for most agricultural policy decisions, are assigned to recipients in several ways: through commodity programs on a select crop basis, by industrial programs that affect processors as opposed to retailers, in allocating funds to specific research operations, and by addressing specific ex/al concerns such as water conservation or food recipient programs. In each case, unique interests have organized around a primary concern with these rather limited issues.[26] Even if organized interest groups such as the National Cattlemen's Association, diversify their policy involvement and become multipurpose organizations, they retain responsibility for public programs that assign benefits to producers of their specific commodity.

Alternatively, public policies can allocate benefits more broadly to recipients who are not grouped by production, industrial, research, or social issue type. For example, farm policy decisions can be aimed at borrowers or at those who store grain in elevators. Neither set of farmers is represented selectively by a single organization that assumes any primary identification with the loan or storage issue. Nor does a single agribusiness organization represent all exporters in any generic sense. When benefits are assigned to farmers, agribusiness, or social reformers this broadly, the players in the policy decision are most likely to include highly diffuse interests or extensive interest-based coalitions.

Given these alternative policy delivery types, what organizational type does government encourage? The answer varies by institution, as can be seen in Table 10.2. Of the forty congressional bills and provisions, twelve affected broad categories of recipients, and only half of these decisions directly affected farmers. In contrast, twenty-eight legislative decisions, or 70 percent, were targeted directly to an assignable constituency represented by at least one nongeneral organization. Sixteen of those involved farmers on a specific commodity basis, not all on price support issues though. The remainder were divided between ex/al interests and programs of concern to business.

This farm commodity orientation was more pronounced for administrative rulings, even when the typically narrower purposes of the regulatory process in general and USDA in particular are taken into account. Of the 115 rulings that assignably target specific constituencies represented by select interests, 87 of these affected farmers on a commodity but not necessarily price support basis. That is, 62 percent of all food-related policies promulgated in the USDA and by executive order were decisions made about commodity policy, whereas only about 9 percent looked more broadly at the generic problems of farmers. As was the case with Congress, most administrative rulings—nearly 90 percent—were made about major commodities such as cattle and wheat rather than the minor crops of potatoes, limes, and guayule. Of the remaining selectively targeted rulings, specific agribusiness problems received more attention than did the combined decisions affecting research, conservation, and food recipient interests.

These data do not indicate that each government decision rewarded the specific interests to which it was targeted in selectively assigning benefits. Determining net winners and losers on the basis of these data is impossible without exploring who wanted what. These data show only the degree to which those who represent highly specific interests have a dominant stake in decisions made about agricultural policy. As noted from the earlier comments and opinions of respondents, much agricultural policy is the limited preserve of farm interests, and the way policy matters are dealt with at its institutional heartland actually perpetu-

TABLE 10.2
THE ASSIGNMENT OF AGRICULTURAL POLICY BENEFITS
BY ORGANIZED CONSTITUENCY, 1980–1985

BENEFITS	CONGRESSIONAL DECISIONS		ADMINISTRATIVE/EXECUTIVE DECISIONS	
	%	Number	%	Number
Nonselectively assigned through				
Farm programs	15	6	8.6	12
Industry programs	7.5	3	4.3	6
Ex/al and rural programs	5	2	4.3	6
Miscellaneous	2.5	1	0.7	1
Subtotal	30	12	17.9	25
Selectively assigned through				
Commodity programs	40	16	62.1	87
Industry support	15	6	11.4	16
Research support	0	0	5.0	7
Conservation programs	2.5	1	0.7	1
Food-recipient programs	12.5	5	2.9	4
Subtotal	70	28	82.1	115
Total	100	40	100	140

ates that situation and gives it a distinctive commodity focus. Public policymakers do indeed tend to treat farm problems through commodity policy.[27] Quite clearly, given their need to respond to the way government structures its decisions, as seen in Chapter 5, there is a reason for the commodity emphasis of all farm groups. Certain policy niches, especially rather specifically defined ones, should be easily maintained when so many issues are brought to an already agreed-upon starting point for further resolution.

However, the data also indicate that not only commodity groups are affected by decisions that selectively assign benefits. Some decisions selectively affect the constituents of more than one organized interest. For example, grazing on public lands is as much a selective issue for the Wilderness Society as it is for the National Cattlemen. One loses when the other wins. And, on a USDA decision to buy canned pork, there are several winners: the pork producers, processors, and institutional users who will receive the commodity. However, neither conflict nor a consensus of support by several players representing diverse interests characterizes the sample decisions that are targeted to a selectively assigned constituency. Usually only a single obvious organized interest, or several related commodity groups, would be expected to be involved. Less than 10 percent of the decisions seem likely to line up opponents, while another 10 percent seem to be of common concern to two or more potential coalition partners.[28] This helps

TABLE 10.3
SELECTIVE COMMODITY PROGRAMS BY TYPE
AND BY INVOLVEMENT OF COMMODITY GROUPS

TYPE OF DECISION	SAMPLE DECISIONS		COMMODITY DECISIONS IN SAMPLE		DECISIONS WITH ACTIVE INVOLVEMENT OF COMMODITY INTERESTS		DECISIONS IN WHICH COMMODITY INTEREST WAS SUCCESSFUL	
	Adm.	Leg.	Adm.	Leg.	Adm.	Leg.	Adm.	Leg.
Price supports	42	10	42	10	28	10	17	8
Trade and marketing	33	6	22	4	12	2	11	2
Taxation	2	1	1	0	0	–	–	–
Subsidized inputs	11	6	3	1	0	1	–	1
Conservation	14	1	6	1	4	1	4	1
Nutrition	10	2	5	0	2	–	2	–
Health and safety	16	2	9	0	3	–	2	–
Total	128	28	88	16	49	14	36	12

NOTE: Adm. = Administrative; Leg. = Legislative.

explain why there are few incentives for formalizing policy coalitions and yielding on an organization's defined private policy identity.

From these data, it appears safe to say that on administrative matters the participants in agriculture consist of a disproportionate number of commodity groups. Cooperatives are also likely to be active players since nearly two-thirds of the commodity decisions involve co-ops as producers/middlemen. Agribusiness trade associations and single firms are much less likely to be involved, but they are the likely partners for both farm and ex/al interests in decisions in which coalition support seems likely. Organizations that represent farmers or retailers in general, such as the American Farm Bureau Federation or the Food Marketing Institute, are likely to become players less frequently than might be expected if they do not get involved in decisions that selectively assign benefits. Far fewer nonselectively assigned issues, both administratively and in Congress, are left for them to broker or exercise leadership on than are covered by the comprehensive networks of established and narrowly focused commodity and single-business interests.

WHAT KINDS OF ISSUES?

Understanding the issues and public programs that assign benefits is extremely

important since the tendency is to think of commodity policy as income maintenance through some form of direct farmer payments or some other price support
mechanism such as the dairy program. That is far from the case. Public policy in
agriculture is promulgated through commodity specific programs in each of the
nine issue categories, except agricultural structure and rural America/poverty,
noted earlier. As can be seen in column 2 of Table 10.3, only 48 percent of commodity-specific administrative decisions provide target prices, guarantee loans,
support fluid milk, make payments to farmers for production losses, or regulate
that process. In Congress, that increases to 62.5 percent.

Commodity-specific programs constitute a majority of the administrative
decisions that selectively assign trade and marketing benefits through grade
standardization, marketing orders, and the like. These programs also account for
substantial percentages of decisions that subsidize inputs such as protection
against gypsy moths for fruit growers, set crop specific conservation practices,
provide nutritional benefits through such policy initiatives as cheese distribution,
and regulate for health and safety through commodity inspections and input restrictions. Health and safety percentages, in fact, would be higher if specific crop
and animal import restrictions made on the basis of disease factors were included.

Congress is less prone than the USDA to govern agriculture on a commodity
basis for these other categories, although most trade policies are set this way. For
the remaining categories, the reluctance of Congress to govern by commodity is
explained partially by the far fewer decisions made by legislators as opposed to
those in the executive branch who implement farm bill policy. Policy decisions are
not only disproportionately structured to encourage the involvement of commodity groups and commodity-based cooperatives; they are structured that way for
most types of decisions.

Who Becomes Actively Involved in What?

Apparently, too much can be made of an organized interest's opportunity to be a
dominant force in agricultural policymaking. As can be seen by comparing col-
umns 2 and 3 of Table 10.3, the expected commodity players become involved actively in most selective legislative decisions on commodity programs (14 of 16)
but only 56 percent of comparable administrative rulings. As would be expected,
most of their attention was directed to price support programs.

The general farm- and protest-style reform organizations, collectively, become involved more frequently than do the commodity groups. At least one organization was involved in each farm program before Congress and in one-half of
the industry programs, especially trade. Moreover, these organizations were active in at least 50 percent of the commodity programs before Congress. Legislatively, they appear more visible than the commodity groups that indicated almost
no involvement in the twelve nonselective policy decisions.

These same diffusely focused organizations reportedly took more active stands on administrative commodity decisions as well, citing sixty-four instances of intervention as opposed to forty-nine for their more numerous commodity counterpart. Again, as in Congress, at least one of the general farm or protest groups became actively involved in all of the nonselective administrative decisions.

Trade associations were represented less frequently than commodity organizations or the other farm groups. There were fewer industry-specific decisions either in Congress or in the administration. However, at least one trade association was active on over 90 percent of these decisions.

But trade associations, for both branches of government, were active on fewer than 10 percent of farm and commodity programs and approximately 25 percent of all the ex/al issues. Commodity organizations were involved less frequently in either industry or ex/al decisions. Their representatives reported only four instances of intervention in fifty-one policy decisions.

These data may not include all participants but are comparable from one type of interest to another and for different membership categories as well. As a result, the data are supportive of those observations that place intensively focused organizations in very narrow policy niches. There appears to be little intervention in either broader issues or those narrow ones of more concern to other organized interests.

Little trade association activity, even by multipurpose organizations, was reported for those policy decisions that have been traditionally dominated by farm groups. The broadly diffuse farm organizations, especially the Farm Bureau and National Farmers Union, and the farm protest groups collectively behave as if the focus of their policy interests were nearly all-encompassing when Congress or the USDA makes an agricultural decision. At least one such organization makes active demands on even narrow issues.

But do these patterns of involvement reflect policy influence? Are the commodity organizations, which are less active on policy decisions than might be expected, also less important in mainstream farm issues than the extensively involved organizations that collectively take positions on considerably more decisions than do their more numerous commodity counterparts? The answer, in both instances, appears to be no.

WHO INITIATES WHAT DECISIONS?

One important measure of interest success is the degree to which interest group proposals become the basis for all agricultural policy decisions. Those who successfully initiate proposals demonstrate the ability not only to influence but ultimately to set the direction for some aspect or issue area of agricultural policy. Given the incremental pace of policy negotiations, the initial proposal is more likely to allocate benefits to clientele than modifications subsequently take away.

By comparing respondent information with that provided by other policymakers, it was possible to identify the institutional origin of 121 of the sample of administrative rulings and all 40 sample congressional decisions. Organized interest groups had initiated 38 (31 percent) of the administrative proposals and 10 (25 percent) of those before Congress. These were disproportionately (28 to 58 percent) regulatory decisions to control the actions or limit the behavior of specific constituents without directly allocating social or economic benefits. Such successfully initiated proposals were usually intended either to reduce the financial or trade costs of regulatory action or otherwise to make product marketing easier. Of the remaining proposals successfully initiated by organized interests, 15 (31 percent) were distributive decisions that directly allocated new financial benefits to the sponsor's members or patrons. Earlier research by Robert H. Salisbury and John P. Heinz on the effects of fragmented demands on the policy process suggests these results.[29] Since each organized agricultural interest makes unique demands of its own, there appear to be few opportunities for following the lead of groups and firms that either want to redistribute program benefits from one constituency to another or allow a particular set of participants the freedom to regulate themselves. The only exception to that general finding was for marketing orders, in which self-regulation remains accepted and institutionally uncontested despite some conflict over the issue.

The difficulty of gaining the endorsement of policymakers for redistributive decisions can be seen in the high rate of rejection for proposals of this type initiated by organized interests.[30] Of twenty-seven decisions in which organized interests reported initiating an alternative or competing proposal that policymakers rejected, twenty-two proposed to alter or made substitutions in the formulas through which direct agricultural policy benefits were distributed at that time. However, in most instances, earlier statutory provisions made some action necessary on these decisions. Consequently, under pressure to rule, administrative officials were credited with initiating most of the commodity price support decisions made in both the USDA and Congress.

Members of Congress, on the other hand, were reported to be the initiators of most of the legislative farm decisions that nonselectively assigned distributive benefits. These also tended to be the type of administrative decision made as a result of requests from legislators. Legislators, in fact, were credited with initiating a few more distributive policy decisions than were interest groups.

These data suggest the degree to which interest groups share influence within the complexities of the policy process. In explaining their role in the initiation of policy decisions, interest representatives explained the process more as one of give-and-take rather than absolutes. The generally agreed-upon pattern of involvement meant waiting to analyze the substance of what the USDA and other administration officials seemed likely to propose. If a group disagreed, or thought that pursuing its disagreements was worthwhile, the next step involved working with policymakers to resolve disputes. Many administratively and legislatively initiated proposals involved some input from organized interests, and respondents agreed that most interest-backed decisions were formulated with prior knowledge that these seemed likely to be accepted by policymakers.

It is instructive, nonetheless, to see whose proposals were most likely to survive the deliberate process at least reasonably intact. Despite the lack of support for their redistributive price support proposals, commodity groups initiated twenty-two (46 percent) of the sample decisions. The highly involved general farm and reform organizations claimed responsibility for only three, or a small 6 percent. So farm groups indeed do seem to be disproportionately advantaged by governmental decisions that are narrow and technical.

Trade associations and other agribusiness interests successfully initiated sixteen decisions, or 33 percent, including several proposals to promote government sales of processed commodities. Ex/al and research institutions had initiated the remaining seven (15 percent) interest-backed decisions, a number identical to the decisions reportedly made on the basis of active requests from foreign governments. Interestingly enough, while there were some instances of commodity or trade organizations jointly initiating a proposal, none were negotiated from the outset by dissimilar groups such as commodity and general farm or agribusiness and ex/al interests.

SUCCESS RATIOS

Organized interests were involved actively in more decisions than those they initiated by more than a five-to-one margin. Although interest representatives claimed to have initiated 48 of the sample decisions, they indicated active involvement in at least 261 instances for these rulings and bills (counting each group active on a single decision).

Their perceptions of success indicate the degree to which the agricultural policy process is highly negotiated, albeit more open to some players than to others. Table 10.4 reports the success ratios of various organizations on policy decisions in which they became actively, rather than just publicly, involved. In almost all instances, group representatives prefaced their remarks by noting that most policy successes were only partial. For example, deficiency payments would be moved in the direction an organization wanted but not set at the desired level. Or, an environmental group would not get the exact language it wanted in a regulatory statute for which it claimed success.

Interest representatives generally report slightly higher success in Congress than for administrative rulings, but rates seem uniformly high for both farm groups and trade associations for decisions that nonselectively and selectively assign benefits. The general farm and reform organizations that claim to be active players on selective decisions are about as likely to feel as successful on the eventual outcome of these issues as are commodity and trade interests. In short, nearly all the established agricultural interests and even some others that are clamoring for recognition, even with only partial successes, perceive governing

TABLE 10.4
SUCCESS RATE OF ORGANIZED INTERESTS WHEN
ACTIVELY INVOLVED IN POLICY DECISIONS

PROGRAM	SUCCESS RATE	
	Administrative	Legislative
Nonselective decisions:		
General farm and		
reform organizations	86.6% (13 of 15)	100 % (14 of 14)
Trade associations	88.8% (16 of 18)	77.7% (7 of 9)
Ex/al interests	0 (0 of 4)	33.3% (2 of 6)
Selective decisions:		
General farm and		
reform organizations	77.4% (41 of 53)	68.9% (20 of 29)
Commodity groups	76.6% (36 of 47)	85.7% (12 of 14)
Trade associations	74.1% (23 of 31)	84.6% (11 of 13)
Ex/al interests	61.1% (11 of 18)	40.0% (6 of 15)

NOTE: Multiple groups may be involved in each decision.

institutions to be responsive to their lobbying when these organizations act as if they have a perceived stake in the final outcome.

In contrast, ex/al interests do not report the same success or, apparently, feel the same degree of institutional responsiveness. Exceptions occur on a few selective regulatory decisions. For the most part, however, they are involved less and win less often than do farm and agribusiness interests.

Those who manage and represent organized interests in agricultural policy follow—or believe they follow—a highly rational strategy in making certain issues the target of their political involvement. Although more intensively organized interests initiate proposals and win with perceived frequency, the more diffusely organized interests must be satisfied by initiating relatively few decisions but being included in negotiations on a great many policy items. Moreover—for some unexplained reason that relates to the negotiated, half-a-loaf nature of policy outcomes—the representatives of the general farm and even agrarian reform groups believe they make at least some gains on most items. Ex/al interests, on the other hand, appear to be equally rational in attempting to move agricultural policymaking decisions beyond the confines of the USDA and congressional agriculture committees in which they may get more satisfying results than they did on the sample decisions.

SUMMARY OBSERVATIONS

This chapter makes clear that, in their active involvement with public policy decisions, organized interests are not the freewheeling entities that some may fear and that many scholars seem to imply. Though their representatives may make many public claims about what they want and whom they support, interest representatives operate under severely restrictive and organizationally self-imposed limitations. These limitations mean that the locus of an organized interest's active public policy involvement is unique from group to group and firm to firm. This situation is a reaction to both the institutional fragmentation of policymaking and the constant need to deal mostly with relatively narrow decisions and provisions that are not intended to produce a cumulatively succinct public policy. Organized interests could not lay aside their differences—even if their representatives wanted to—in order to initiate or even support a comprehensive and internally consistent farm policy, let alone a comprehensive agricultural policy on farming, food, fiber, and trade.

When studied individually as a vaguely linked representational system, agricultural interests give new meaning to Bauer, de Sola Pool, and Dexter's contention that organized interests are better understood in terms of what factors limit their influence rather than their ability to pull out all the stops in a search for policy dominance. While agricultural interest groups would like to find the formula for dominance and while they constantly organize and develop better tactics to do so, all they can seriously hope for is a limited place in the policy process. Political observers keep lapsing into the mistake of adding up the lobbying resources of organized interests and concluding that group and firm representatives can buy the whole policy process. To borrow an analogy, organized interests, no matter how resourceful, can use only their resources to buy a part of the Washington real estate. Just as no organization can make a rational choice to buy all the property on K Street or in Georgetown, no rational interest wants or is able to spend what it takes to seize control of the national agricultural policy process. That property is made too expensive, first of all, by the expanding universe of organizations in an increasingly interest group-oriented society and, second, by the number of rather autonomous policymakers who are responding to a variety of influence agents beyond private agricultural interests. Equally important, with their short-term focus on policy and their internal pressures to deliver immediately useful benefits to those who pay their bills, why would any private interest representatives want to be able to influence every facet of agricultural policy?

The answer to the question of why an interest group might want to control all of agricultural policy may rest on some human tendency toward dominance, a fear articulated at the heart of the agrarian protest claims of conspiracy. But interest groups, within their loosely structured representational system and fragmented policymaking environment, do not behave that way; organized interests make many demands but most are public posturing. Entrepreneurship, while it explains the movement of interests into new and expanded policy ground, is limited severely by the perceptions of those who share the political environment and by the actual policy responsibilities of the organization. If an entrepreneur wants to expand his or her involvement for ego gratification, he or she must do it beyond the confining employment of a single interest. Finally, representatives of organized interests behave as if they were transactional actors who share policymaking with others, most frequently those in policymaking institutions that emphasize common problems. Organizational interests initiate very little, especially for some types of policy decisions; their representatives behave primarily as one negotiating partner in the eventual resolution of decisions that are amazingly short-term in that they will soon need to be made again. Perhaps most surprisingly, these individuals feel satisfied enough with the outcome of negotiated decisions to claim high rates of policy success.

Even if lobbyists are not useless, as Heinz Eulau once questioned and many business executives adamantly believe, they cannot be relied upon to control agricultural policy.[31] The next chapter, on the politics of the 1985 farm bill, reinforces this point.

11

Private Interests and
the Food Security Act of 1985

Despite the high incidence of their satisfaction with agricultural policy decisions, little evidence exists to show that organized interests dominate policymaking. As seen in Chapter 10, even the most established interests operate in a setting in which key policy initiatives originate from many sources. It would be necessary to reject any model of agricultural policymaking that portrayed government as little more than the reactive element in a political process in which narrow special interests defined acceptable policy alternatives.

This chapter provides evidence in support of an alternative explanation that still attributes considerable policy importance to private interests. In this explanation, policymakers must be seen as facing numerous problems in deciding upon appropriate responses. Private interests gain influence precisely because organizations and coalitions capture some part of the attention being directed to problem solving. During the deliberations over the 1985 farm bill, a large number of individual interests were able to capture this attention and hold it long enough that certain ideas and goals were identified as essential to the act.[1]

If the policy arena is sufficiently complex, as it is in agriculture, a large number of diverse organized interests can gain policy influence and feel successful. They win, in this respect, because their specific policy niches provide each of them advantages that other policy participants lack. The biggest of these advantages is the perceived legitimacy by policymakers as the appropriate spokespersons on specific types of policies. As the respondents cited earlier suggest, interest groups and firms become influential in agricultural policy because their representatives respond quickly to political conditions and events on their policy turf. Sometimes the part of the problem being resolved is one that the private interest is simply expected to understand thoroughly and be able to interpret for policymakers. Then, like the commodity organizations and agricultural experts, these interests are looked to for advice. More frequently, however, organized interests capture attention by defining their specific understanding of agriculture conditions as germane to unraveling policy circumstances. Both agrarian protestors and trade-dependent associations are relied on to interpret, respectively, farm and international conditions because they have successfully convinced policymakers of their own organizational proximity to what goes on outside Washington. In this way, such organizations often try to redefine policy problems as those that relate to their experiences. For example, some agribusinesses have attempted to redefine the farm crisis as a trade crisis.

This view of organized interests, based on transactional policy relationships and the power of usable information, suggests that private influence is real but highly situational in terms of ongoing social, economic, and political events. Influence, in other words, is neither an absolute nor is it necessarily enduring. Rather, it rests on how appropriate an interest's resources, tactics, and organizational type are to the problem and proposals being considered.

For example, as seen in Chapter 10, commodity organizations have gained reputations for power in agriculture even though they are no less reactive than other interests, become involved in policy decisions less often than might be expected, and claim no higher policy success ratios than other farm and agribusiness interests.[2] Situationally, however, commodity groups do benefit more than any others from the procedural aspects of agricultural policymaking in Congress and the USDA. They do so because of the commodity orientation of most farm legislation, the agricultural subcommittees in Congress, and their own organizational focus on attainable and selectively assigned farmer benefits.

This chapter contends that much of the influence of agricultural interests must be understood in the context of cyclical farm bill legislation. Of course, as seen in Chapter 2, the fragmentation of agricultural policy among many institutions provides other avenues of obtaining private policy goals. Nonetheless, the farm bill occupies center stage because it gives so many interests a chance to make policy gains.[3] Because organizations take advantage of the actions and behavior of other private interests as well as of policy conditions, influence tends to be segmented rather than shared; and, predictably, successes tend to be limited, based on political dominance of a policy decision.

FARM BILL POLITICS

The farm bill was conceived in contradiction in 1933. Although that innate contradiction is inherently sensible and probably unavoidable in that farmers indeed do produce on a commodity-by-commodity basis, the procedural approach of dealing with the comprehensive problems of agriculture on a provision-by-provision basis has led to confusing expectations about how government should manage its relationships with agriculture.

The Agricultural Adjustment Act of 1933, which eventually begat the cyclical farm bill process of renewable legislation, was designed in a depression era of massive surplus in order to restore farm purchasing power. Parity was to be pursued on an agricultural-wide scale by empowering the secretary of agriculture to develop and implement a total strategy involving nearly all aspects of the agricultural community. The secretary was authorized to: make acreage reduction agreements for basic crops, use direct producer payments to secure production cutbacks, arrange agreements with agricultural middlemen to regulate marketing, license middlemen to control unfair handling practices, determine taxes for processed goods to support these regulatory efforts, manage adjustment operations in other ways necessary for market expansion and surplus reduction, and ensure that consumer prices did not go beyond the parity level. The 1933 Act and 1934–35 amendments designated barley, cattle, corn, cotton, flax, hogs, milk, peanuts, potatoes, rye, sorghum, tobacco, and wheat as basic commodities and the units for farm policy management.[4]

After some immediate crises precipitated by court rulings, farm bill politics settled into a routine for dealing with the major national issues of food and fiber production and use. Goals ranged from World War II plans to keep production high for the war effort to 1973 initiatives to capture the expanding world trade market by providing incentives for increased planting. The chronic problem to which policymakers had to return, however, was one of commodity surplus and overproduction resulting in low farm prices. Environmental use and conservation always came up as issues of national consequence within the context of manipulating commodity supplies. Under these circumstances, it was inescapable that farm, food, and fiber policy came to mean commercial commodity policy. The situation was no different in 1973 when worldwide demand for feed grains was high. Agricultural legislation was predicated on the need to encourage production and remove costly programs that curtailed production. These plans were selectively assigned commercial commodity programs, not the alternative of general economic or trade mechanism policies. In the 1930s, for example, one of the surplus/low-price problems was a sow and pig problem, with its solution inherent in the reduction of hog supplies. In the marketing-oriented 1980s that same problem of oversupply led to government support for a pork promotion program.

This situation eventually engendered serious criticism. Theodore Lowi articulated one of the concerns of a generation of political scientists and agricultural economists who studied the agricultural policy process when he described the renewable farm bills as little more than ten titles purporting to be a single bill.[5] As Lowi saw the farm bill process, it was unable to deal with any more than the narrow issues of the narrowest farm interests. It reflected neither comprehensive economic planning nor attention to the food-and-fiber-related social problems of the world or the nation.

Lowi's position, however, represents only one side of the farm bill dilemma, the one that owes itself to the bill's commodity-specific heritage. There has long been a fear that commodity-based policy decisions serve no national farm policy purpose. Political observers predicted an impasse in farm bill legislation in the 1960s as commodity groups proliferated in Washington affairs, leaving the divided general farm groups without the ability to broker the bill on the basis of some central allocation of commodity benefits.[6] In the 1980s, the assumption remains that in order to pass, the farm bill must deal with those national problems of comprehensive food and fiber need around which the bill originated.

This belief that a farm bill can be too narrow has led to a hectic search for national policy relevance and the oft-repeated fear that this will be "the last farm bill" that will pass on its now nearly four-decade-old four-to-five-year cycle.[7] Since 1973, farm bills have passed as Congress put together broadly supportive urban-rural coalitions of first labor-oriented and then consumer-conscious members who backed their farm state colleagues. Farm state legislators have traded votes on minimum wage laws and the Consumer Protection Agency for farm bill support.[8] To popularize farm bills and broaden their purpose, food stamps, food aid, and consumer provisions have been incorporated. Although the agriculture committee leadership has been partially responsible for these programmatic additions and tradeoffs, coalitions of southern and midwestern legislators have contributed to the process of building a winning majority through their own efforts to allocate benefits fairly and satisfactorily by region.[9] These regional coalitions are commodity-oriented, with southern legislators concerned with cotton, rice, peanuts, and—even though it recently has been given separate legislative status in another bill—tobacco. Midwestern legislators speak for wheat and feed grains. Because of regional diversification, active sugar and milk proponents come from a wider range of states. This disjointed array of supporters, in and out of committee and perceived to be solely responsible for the ability to reformulate each new farm bill, understandably has been viewed as an unstable foundation for farm legislation. In 1976, as a corrective measure aimed at securing more ongoing congressional support, the Young Executives Committee of the Department of Agriculture recommended that remedial action be taken within the department to broaden the primary clientele "to incorporate effectively the interest of low-income consumers . . . and other groups of society affected by (USDA) functions."[10]

Although the farm bill process has gained a reputation for its narrow commodity emphasis, the legislation and its surrounding circumstances have not been without a broader focus as policymakers have attempted to maintain its original intent as a national policy. This attempt to perpetuate the contradictions of farm bill procedures has left both policymakers and political observers somewhat unclear as to what the farm bill is really all about. This has left the act a catchall policy document, one that many agricultural interests wait for in an attempt to insert a single provision that will generate little conflict in the noise and rush of passing omnibus legislation. Clearly, the renewable farm bill has become more than ten distinct provisions masquerading as a whole. Beginning with the Food and Agriculture Act of 1977, the farm bill has brought together a great many types of programs beyond just commodity price supports.[11] Trade, conservation, credit, food distribution, research, extension education, nutrition, marketing, inspection, animal welfare, and constituent participation provisions also are likely to be included in any act. Yet, despite the attention, the farm bill process has not become a public forum for debating, in any integrated or cumulative way, the major policy problems of a changing agriculture. Ross B. Talbot, a political scientist who has been involved more continually in agricultural policy than anyone else in that discipline, summarized the dilemma of the farm bill as he argued for more attention to world food needs.[12] Talbot noted that though food aid programs such as Food for Peace occupy a place of some prominence within the farm bill, an important and related issue such as U.S. involvement with world food organizations will not become the topic of enough attention to resolve the problems of U.S. participation. Time is too short, organized interests that would be important to these debates have other priorities, and the agenda of pressing issues is too great for policymakers to force attention to this single, nonchampioned issue.

Talbot's example emphasizes the need to understand more about private influence over the farm bill. While it is understood as a massive, often frustrating example of congressional coalition building, what role do organized interests play in its assembly? Because of the importance of gaining selectively assigned commodity benefits and the capacity of organized interests to maintain a defensive advantage, do they inhibit policymakers from comprehensive policy reform by their dominant commodity focus? While earlier evidence indicates that organized interests are not the initiators that provide the sole momentum for the decisional process, are they—as individual organizations or as a collective system of representation that come together for this single occasion—impediments to change? The answer to both questions appears to be yes.

THE SETTING FOR 1985 LEGISLATION

The conditions of American agriculture, as described earlier in the book, were characterized by recurring surplus and low commodity prices. Farm economic conditions preceding the 1985 farm bill were worse than usual, though, because of the rapid farm expansion of the 1970s followed by escalating production costs and debt and a shrinking foreign market for U.S. farm products. Most importantly, these agricultural problems were negatively affected by macro-economic problems that had stalled both U.S. and international economies. The combined effects of the 1979 Federal Monetary Act, with its anti-inflation goals, and the 1981 Reagan-administration-inspired tax cut led to even higher interest rates, a highly valued U.S. currency on the international market, a general recession in the early 1980s, and an unavoidable declining demand for U.S. food and fiber. Operators of large and middle-sized farms were being forced to leave agriculture, just as their smaller-scale peers had done since the depression of the 1930s. With large increases in production capacity promised for the future and with uncertain prospects for U.S. agriculture outcompeting foreign exporters, American agriculture was thought by most policymakers and interest group representatives to be in a transitional period in which many more producers would be forced from the market.[13] The situation was not just viewed as a normal adjustment downward in the number of farmers as land was acquired by neighbors and kept in production. Acreage, investment, and outlay for farm inputs were all predicted to decline in the forseeable future.

It was in this economic and agricultural context, with its accelerated sense of urgency, that participants prepared for the 1985 farm bill with the generally articulated belief that existing policies were either outmoded in dealing with changing agricultural conditions or that they actually worsened the situation. It was difficult to find any participants who did not express the opinion that the time for major agricultural policy reform was at hand.[14] At the same time, however, many economic policy planners believed that any policy changes in 1985 would create only short-term adjustments in demand because policymakers could not deal with macro-economic problems in any farm bill. So, while the time for reform seemed appropriate in terms of both need and zeal, there was little faith that reform would be accompanied by the immediate policy benefits that either farmers or industry wanted.

Reform had an important political context that affected interest behavior and demands as well. In that regard, the political dynamics of the Food Security Act are also different from earlier farm bills. Seven political factors were perceived by representatives of organized interests to be critically important in shaping the farm bill. First, the farm crisis became evident nationally as farm foreclosures and bank failures called attention to a general economic deterioration in rural America. The spread of economic hardship and national concern with the problem threatened a growing political backlash against policymakers who became increasingly unwilling to offend farmers. Second, congressional Republicans became fearful that farm and rural economic conditions would spell personal electoral disaster in the 1986 elections if not successfully addressed. With twenty-two Republicans up for election, over half of whom were from important farm states, the Senate appeared to be especially vulnerable to partisan shift. Thus, there was little Republican support for the Republican president. Third, despite congressional concerns, the Reagan administration's interest in early farm bill deliberations took the form of rhetoric, challenging growing costs of farm programs and calling for the elimination of most price supports. Conflict between White House advisers, the Office of Management and Budget, and the USDA led to a very late submission of a farm bill. When it arrived in Congress, USDA officials were left to defend a proposal that in the wake of the farm crisis struck most members as what one called "at best a joke and at worst a philosophical tract." It was not, at any rate, a proposal that allowed USDA experts much leadership in the farm bill process. It did leave interest representatives greater latitude in promoting their own initiatives.

Despite this general congressional concern for aiding farmers, a fourth political factor was no less evident. The federal budget was seen as excessively large and constantly growing. Secretary of Agriculture John Block's repeated statement that a balanced budget would be the most productive farm bill found considerable sympathy among legislators concerned with agriculture's high-interest rate problem and its escalating effect on farm debt. The nearly fivefold (from $4 billion in fiscal year 1981 to nearly $19 billion in fiscal year 1983) increase in commodity price support outlays made it appear "that farm program costs were out of control and was cause for alarm."[15] Every organization, even of parity supporters, recognized the budget as a limiting factor. Only a few private representatives understood a related phenomenon, the reluctance of many policymakers to back policy reform when it was unlikely to alter a farm or a trade crisis immediately.

Budget deficits had triggered a sixth political factor in what passed eventually as the Gramm-Rudman-Hollings Act, which mandated across-the-board cuts in federal programs if deficit reduction targets were not made. Policymakers had not only to consider the deficit but also to write farm bill programs with the impact of possible cuts and ways of avoiding those cuts in mind. Other legislative problems constituted a seventh contextual factor for the farm bill. Both FIFRA and cargo preference legislation were before Congress, spurring both allegiances and potential pay backs among those who had become friends or opponents of the two agriculturally important bills. The Superfund for the Environmental Protection Agency and the tax reform proposals were also important but less immediate agenda items that affected legislative relationships on agricultural issues. The shifting alliances—both within Congress and among concerned interest groups representing the maritime industry, labor, various agribusinesses, and environmentalists—were related substantively to the content of the legislation and tied politically to the personalities of key participants. For example, Senator Ted Stevens of Alaska, privately still bitter after losing the Senate majority leadership race to Robert Dole by three votes after the 1984 election, chaired the Merchant Marine Subcommittee, with its responsibilities for the maritime industry and its opportunity to consider and finally recommend rejection of the agriculture committee's cargo preference bill. Assisted by loyal lobbyists and friends who represented home state interests, Stevens helped drag on the cargo preference dispute over which Dole presided. Dole's inability to bring closure on that bill with any degree of quickness led many to doubt his leadership capacity on the future resolution of the pending farm bill. He had let too much anger and conflict divide agricultural policymakers and lobbyists on what many saw as a relatively minor matter. Collectively, those other issues created among legislators a greater urgency to cooperate in passing a farm crisis farm bill.

As a whole, these political conditions produced an environment in which, by 1985, farmer assistance programs were generally viewed as a necessary evil, but there was no agreement as to what was the best. Moreover, the White House and Congress were divided as USDA officials sat idly by, useful as legislative analysts but largely inconsequential as policy leaders. Within Congress, where the bill eventually would have to originate and be resolved, leadership was being questioned. Loyalties and policy preferences were no less divided by personal, partisan, and ideological differences. The representatives of agricultural interests were forced to ascertain their goals and develop lobbying strategies while reacting to these unfolding events and the increasing recognition that there could be no consensus among policymakers about what directions agricultural programs should take during what could have been a period of substantial reform.

INITIAL STAGES:
ACADEMICS AND AGRARIANS

The strategies of private interests for influencing the 1985 farm bill debates originated many months prior to formal congressional deliberations. Since policymakers value factual information about policy conditions as means for reducing the uncertainty of risky decisions, and because the 1985 farm bill promised to be an enterprise fraught with considerable political and policy risks, it behooved no one to enter the decisional process unprepared. Constituents were being mobilized, political funds were being raised, consultants were being activated, and extensive policy analysis was underway, all orchestrated by interest representatives who wanted to promote their own interpretations of agricultural conditions. Because of widespread perceptions regarding agriculture's critical status and beliefs about the appropriate timing for policy reform, those who wanted to become players in farm bill politics began to marshal their potential resources early to establish the credibility of their claims. For two sets of private interests that are not part of the established Washington lobby but that had become expected players of some importance, this meant that preparatory discussions started in 1983 and produced high levels of activity throughout 1984.

The most detailed and analytical responses to the pending legislation came from individuals and institutions that are often overlooked as private interests but that, nonetheless, are important for the ideas they generate and bring to the policy process.[16] Lumped together, they can best be described as academic agricultural experts. What was unusual about this loosely knit collection of professors, foundation representatives, government-attached agricultural analysts, and some technician-style private consultants was their common opposition to the nonmarket orientation of existing farm policy. While there was little or no agreement on the specifics of reform, there was a strong and important consensus on the need for reducing farm program costs and moving to an internationally competitive U.S. agriculture as the only acceptable means for minimizing surplus. Even former champions of price support programs such as former Kennedy administration adviser Willard Cochrane came forward and questioned the continued usefulness of commodity price support programs.[17]

In addition to their common conviction that present policy operated to the severe detriment of American agriculture in an international setting, other nonagricultural factors motivated the experts' policy involvement. The appearance that the time was right for major legislative reform made it an equally appropriate time for those with the greatest analytical skills to engage in their own form of entrepreneurship, for consultants to advertise their talents and services, for foundations and universities to demonstrate their policy relevance as consequential public forums, and for the financially supportive agribusiness and commercial firms that funded policy conferences and research to demonstrate their own commitment to better-informed public policymaking.

For the most part, these players neither intended to be nor considered themselves part of a lobbying effort. Nonetheless, the experts wanted to influence farm legislation and were willing to represent their ideas forcefully. Also, the individual players were participating under the auspices of institutions that felt a need to secure and continue to occupy credible positions in the policy process. The Farm Foundation, for example, has long been considered one of the facilitating forces "behind farm policy."[18] Foremost among newer activists were the National Center for Food and Agricultural Policy of Resources for the Future and Monsanto Company, an agribusiness firm that brought experts together. The emerging National Center sought to establish its reputation in developing agricultural leaders and providing reputable policy analysis. Firms such as Monsanto with its *Outreach* conferences were stressing greater company involvement in future agricultural policymaking. It was this philosophical consensus that provided a common interest and the resulting force and momentum of a lobby.

How did this loosely knit alliance of individual interests attempt to influence the 1985 farm bill? With a few individual exceptions, groups did not work on congressional offices, draft legislative proposals, activate constituents, or detail political strategies for passing related bills. Instead, they addressed the information needs of policymaking through well-funded, nationally organized conferences and the publication of relevant research proposals.[19] From May 1984 through January 1985, well-advertised conferences were held at the rate of nearly one per month. In between these meetings, participants gathered to discuss policy issues at numerous other conferences sponsored by the Extension Service of USDA, professional associations of agricultural economists, and several state governments. All of this provided the experts with the opportunity to articulate the dilemmas of agriculture regarding production and trade problems, define policy solutions in terms of free-market values, and, for at least the time being, dominate the only visible public platform upon which agricultural and food questions were being argued within policymaking circles.

If academics provided the policy messages of 1984 and early 1985, a decidedly different message was being prepared for Congress and the public in rural states and districts well beyond Washington's Interstate Beltway. Agrarian protesters planned to deliver this message back home in legislative districts, directly through constituents who promised to vote. As similar groups had done since 1977, they planned to use the help of empathetic and curious media. As noted in Chapter 4, farm protest groups were active in 1983 and throughout 1984, organizing local residents and preparing organized demands for 1985 legislation. As far as farm protest activists were concerned, the government could worry about exports and trade problems all it wanted to as long as immediate steps were taken to enhance farm income and reduce problems associated with rising farm debt, regardless of the considerable conflict in meeting these diverse policy objectives.

As the protest activity of 1984 grew out of the grassroots remnants of the American Agriculture Movement, the center of protest shifted north and east from Colorado, Kansas, Oklahoma, and Texas to include Iowa, Minnesota, Missouri, Nebraska, and Wisconsin, to blanket most of middle America. Activism of some sort, with associated protests, was evident in nearly two-thirds of the states. Although the alliance of small, autonomous groups of farmers, clerics, and other social activists was inherently unstable, its visibility nevertheless gave agriculture an imposing public presence that politicians could not ignore.

Despite academics' hopes that the force of their analytical work would keep their views in the political mainstream in 1985, the agrarian protesters upstaged them with old-fashioned pressure politics as farm conditions worsened throughout the spring. The consequences were evident by midyear. As foreclosures and rural business closings increased, several farm suicides attracted national attention. Pointing to these results, many farm state legislators became committed to policy approaches articulated by protest organizers. Senator Tom Harkin (D-Iowa) and other midwestern legislators became advocates for the Farm Policy Reform Act and, later, the referendum for mandatory production supply controls as these proposals were put together from the Hightower statements by the Iowa Farm Unity Coalition, Minnesota Groundswell, the Nebraska Farm Crisis Committee, and the Wisconsin Farm Unity Alliance. Other farm state legislators found it personally unpalatable but politically impossible to do anything but insist on high levels of direct financial support for their producers. One uncomfortable congressman complained that, even when he spoke of what he considered supportive legislation for the farm protesters, constituent reactions and pressure made it "feel like a foreign country back home." By the end of the summer of 1985, farm protests had exposed large numbers of legislators directly to *the* farm problem. The remainder could not miss it on television network news and from their colleagues' stories. This activism and attention—backed as it was by incidents of tragedy and a growing number of financial failures—had conditioned the public to expect agricultural policy reform to address what farm activists had defined as the farm crisis.

REGROUPING: GENERAL FARM ORGANIZATIONS

Professors and protesters were preparing for the farm bill long before many Washington-based farm and food lobbyists had had much time to make plans. Other legislation and administrative rulings took much of the attention of the lobbyists. The American Farm Bureau Federation (AFBF) could not afford to delay its preparation, however. AFBF officials and staff perceived their organization to be troubled by three problems. First, since the more intensively directed commodity organizations were being viewed increasingly as more capable of affecting major legislation than was the Farm Bureau, less and less attention was given by policymakers to the AFBF. Second, farm protest groups were populated by large numbers of Farm Bureau members who wanted both types of organizations to be responsive to their financial troubles. The lack of common goals between the protesters and the AFBF created membership difficulties. Third, probably as a result of the above problems, there had been increasing conflicts between the state farm bureaus as well as between the states and the national organization. The Alabama Farm Bureau operated independently, and a few other state farm bureaus often lobbied for their own policy preferences, especially on commodity programs. None of this enhanced the image or the influence of the nation's largest farm organization during a period of declining membership.

In a rather bold attempt to restore member loyalty and regain some measure of policy leadership, some Farm Bureau leaders held a series of meetings with commodity organizations that were insisting on maintaining high levels of price supports even in the face of declining exports. The intent, in a manner reminiscent of the National Farm Coalition, was to find common ground for a mutually acceptable farm bill proposal. Agreement was set rather tenuously on most midwestern crops, with the principle exception of dairy. That agreement, however, was a modest victory for the Farm Bureau. Attendant discussion between groups produced little or no willingness to experiment with income maintenance programs that might offer alternatives to the high levels of deficiency payments and the type of loan guarantees found in the 1981 farm bill. As a consequence, the long-time leader of the free-market philosophy in agriculture lost any opportunity to stand firmly in line with either the academic critics of farm programs or the Reagan administration. Still, the organization could more credibly demonstrate some important support for the economic plight of many of its members after participating in these negotiations.

The Farm Bureau did not emerge as a policy leader, however. Little agreement was reached on provisions of the farm bill beyond some basic price support levels. In addition, the Farm Bureau proposal that was eventually drafted encountered difficulty in finding a sponsor willing to be politically positioned with the AFBF, and, when the proposal was sponsored, it was lost in an extensive array of bills introduced in Congress by members who wanted some share of the credit for attending to the farm crisis.

The other general farm organizations fared little better. The National Farmers Union and National Farmers Organization (NFO) found themselves left as the main defenders of the income maintenance remnants of New Deal farm policy by mid-1984. NFO lobbyist Chuck Frazier made a special point of attending and addressing some of the academic reform conferences and championing what was not accepted by him or privately by many other lobbyists as the lost cause of traditional commodity price support programs. Cy Carpenter, president of the Farmers Union, was embroiled in conflict with the farm crisis organizations and grassroots activists from the AAM over the NFU's policy position. The staff of both organizations worked, often together, to formulate policy responses throughout what was rapidly becoming a period of controversy among the advocates of a strong maintenance-directed stance on improving farm incomes. Staff reactions were motivated by fear. At worst, the mandatory production control views of the emerging agrarian protest coalition, centered around the Farm Policy Reform Act, could become the sole liberal alternative to a market-oriented farm bill. The alternative would in turn be identified as an extremist approach, lose, and then open the door to a dismantling of traditional price supports as called for earlier by President Reagan and former Budget Director David Stockman. The NFU and the NFO, in such an event, would forfeit their credibility and status as the long-standing interest group spokespersons for liberal agricultural policy.

The NFU was in a particularly difficult situation because 1985 brought considerable pressure from old liberal allies and legislative supporters who were the protesters' newfound friends in Congress and the Democratic Party. Agrarian protesters were much more useful to what was emerging as a contemporary populist strategy to energize the electorally sagging Democratic Party. To avoid conflict and still lower the potential for commodity program pragmatic losses, the Farmers Union announced its support for mandatory supply controls and publicly spoke—or postured—on behalf of those amendments and bills that contained the appropriate provisions. At the same time, it became an open secret to policymakers and lobbyists who opposed controls that the NFU and the NFO preferred modifications in existing programs for which their representatives also lobbied.

These two liberal groups were not alone in their difficult middle ground. As the farm bill process continued, the organizational leadership of the American Agriculture Movement, Inc. (AAM), would come to have similar conflicts between policy beliefs and their actual policy demands. Although the Bedell, Alexander, Harkin, and Zorinski production cutback proposals were popularly identified as AAM provisions, there was frequent mention by group leaders that they considered mandatory controls as only the best of many bad bills. Despite AAM offices serving as Washington headquarters and primary information source to the fragmented coalition of grassroots protest organizations, many of its leading activists could not fully accept a bill that was the handiwork of Jim Hightower and four state crisis committees rather than of the national movement.

It was this kind of organizational behavior that led members of Congress to begin to conclude as early as 1984 that the forthcoming farm bill would once again be one of incremental changes rather than broad reform. Its members could hardly deny the impact of the well-publicized farm crisis. There were worthwhile economic plans from the academics but absolutely no information coming from them on precisely how to translate a market-oriented approach to the political agenda. Lobbyists brought forth competing proposals without effective challenge by pointing to the self-imposed political isolation of the academics—"those people don't have to be practical."

Using free-market arguments, the Reagan administration spoke of its forthcoming proposals with budget and program cuts that were so extensive that they were denounced as ludicrous by legislators about to face the 1986 elections. The administration's major farm group ally offered a bill that looked to many as if it had been drafted by the NFU rather than by the AFBF. Other interested farm groups that were legitimately able to offer a comprehensive proposal were having trouble agreeing, had little commitment to the major changes that they put forward, and seemed to dislike the most vocal of the activists whom they represented. On the whole, by the time hearings on the legislation began, Congress was not getting much usable information or any coalition support from those private interests that might have been useful to some kind of comprehensive policy reform. In the end, these organizations did little more than the expected—communicate to Congress that they could live with commodity programs that contributed substantially to farm income.

MAINTAINING THE HIGH GROUND:
COMMODITY GROUPS

Specific proposals and supportive data were plentiful, however. But they would be of no help in fostering the kind of reform so often mentioned in 1983 and 1984, whether in a market-oriented or production control direction. The commodity organizations, knowing full well that they would have plenty of opportunities to develop program initiatives with the commodity subcommittees, entered 1985 without the elaborate preparation of many of the other private interests. But they had given a great deal of forethought and in some cases prepared the most sophisticated quantitative analysis of any of the involved interests to those selec-

tive aspects of farm policy needs. Most of the Washington-based commodity lobbies had been watching the mixed signals of 1984 with considerable attention,
and the need for living with the cost-reducing constraints of the budget deficit
was their biggest concern. A great amount of internal organizational effort, as a
result, was being directed by their staffs toward reaching intracommodity agreements with financially needy producers on those major provisions that would directly affect single crops and livestock industries.

To some extent, these groups paid attention to the whole of agricultural policy. Some, such as the National Cattlemen's Association and the National Broiler
Council, did so with more active involvement than others. Both were attracted
philosophically to the free-trade issue in the expressed hope of export expansion.
On a more practical basis, nonmarket commodity assistance programs had
brought dairy cattle to slaughterhouses in competition with beef cattle. It also
created higher prices for feed grain for chickens. Also, on a practical note, these
self-identified free traders limited their market-oriented enthusiasm to only
those circumstances in which foreign beef and chicken imports currently were
not a problem, reserving the right to alter their position if conditions were to
change. Much of their involvement was perceived to be little more than posturing
toward reform while concentrating on their single-commodity problems.

However, quite predictably, most major commodity representatives were
interested primarily in price policy as it assisted their producer members. While
organizations such as the National Association of Wheat Growers examined the
prospects of experimenting with marketing loans as advanced by the National
Council of Farmer Cooperatives, other groups felt such concepts too difficult to
explain. As a result, only cotton and rice interests, with their co-op constituencies, became serious about the marketing loan strategy of letting farmers sell on
open markets and having government directly pay the difference between the
market price and the guaranteed loan amount. Policymakers, the representatives
of most interests believed, were in an environment in which so many alternative
proposals were being introduced that confusion might kill the entire farm bill.
Rather than risk that, those groups whose members received commodity benefits considered three factors: how to maximize deficiency payments and keep
members happy as well as financially afloat, how to bring commodity prices down
in order to enhance exports, and how to discourage budget-deficit-producing and
price-destabilizing surpluses. As commodity officials grappled with these concerns, alternatives to present price programs became harder to handle and propose because the objectives of high price supports and lower world prices
conflicted poignantly.

Since each commodity program varied in procedure, language, and operation, each group used its specialized knowledge to propose and negotiate modifications. Other commodities were left alone unless it was possible to propose ways
of diverting one commodity's program benefits to another. Some coalitions were
formed, such as between long-allied wheat and corn interests or between sugar

and dairy interests, which had been under attack by user coalitions, but the focus of these agreements was always on specific commodity price support provisions.

Additional commodity issues were injected into the farm bill debates whenever the producer groups could do so, further complicating the act but also winning it supporters. The National Pork Producers eventually decided on a hog checkoff as their priority in the bill. Soon the Cattlemen wanted one for beef. The American Soybean Association, traditionally a group opposed to farm programs, wanted one-year crop payments for bean producers in return for lowered loan rates. Sunflower producers, mostly through legislative initiatives, gained direct payments in the Senate. Despite congressional agreements at the onset of 1985 to avoid farm bill action on marketing orders, additional order provisions were worked into the bill by those representing several different interests. In addition, as was the case with other 1980–85 agricultural policy decisions, many of the farm bill's final trade provisions were drafted to provide commodity-specific benefits as well.

Throughout the farm bill process, the commodity groups demonstrated a considerable ability to influence individual provisions and an almost total incapacity to be agents for serious policy reform. Loan rates, under which farmers sold their products to government at a guaranteed price, were lowered in the bill in a concession to market-oriented policy. Export assistance was enhanced. But the greatest contribution of the commodity lobby was in generally holding the line and actually getting government to spend more on major provisions of a bill that none of the group's representatives had favored several months earlier—income assistance through such programs as target price deficiency payments.

IN AND OUT:
AGRIBUSINESS REFORMERS

Early in the process, when feeling ran high that the "Reagan climate" would prevail, it was felt that the 1985 farm bill would be the first to reflect in any major way the views of business rather than farmers. From public positions taken on commodity price support programs, it appeared evident that there was considerable agribusiness dissatisfaction with present policy. Moreover, agricultural experts were rarely united around ideas that coincided so well with basic business values as they were in the months prior to 1985. In addition, agribusiness and industrial leaders, both publicly and among themselves, frequently spoke of a renewed need to be more effective in influencing agricultural policy. There seemed to be a new awareness among business firm CEOs in particular about how directly business profits were influenced by government programs.

This opposition was not united in purpose. Land diversion programs, especially since the consciousness-raising effects of PIK-induced planting reductions, were under attack from farm input suppliers. Sugar and dairy programs served to mobilize user coalitions to reduce politically inflated prices. Peanut and honey programs were similarly targeted for elimination. Traders and shippers wanted reductions in programs that kept domestic prices high and inhibited exports. It appeared as if the agribusiness community believed that any farm policy that interfered with market conditions was bad for business.

While that belief may have been held by many interests, it was not more easily translated into action in 1985 than it had been for the previous three decades. There were several reasons, none of them new and all of them relevant to business leaders as they considered their immediate involvement in farm politics. Lobbying uses both money and time that might go to other enterprises. For the farm bill, with its entrenched producer clientele and its structural features centered around commodity provisions and commodity decision makers in Congress, the costs of effective lobbying will always be high, and returns uncertain. Victory can hardly be assured in advance, and even in victory there can be no specific guarantees as to financial savings and profits. Compromises, bargaining, and negotiations make predictions on the financial implications of policy outcomes difficult. All this, along with the healthy distrust that many business executives feel toward the political process, caused many agribusiness firms and trade associations to proceed slowly, even after these organizations had assumed a vocal public posture about their concern for farm legislation. In short, agribusiness did not end up putting its collective lobbying money where its initial posturing indicated that it would go.

The political dynamics surrounding the farm bill affected the plans of business and industry as well. Those producer organizations were unwilling to promote substantive reform after their leaders had spoken so disparagingly of farm programs. This removed much of the potential coalition base between farm and business that was felt necessary for agribusiness to affect change. In the event of open controversy, the farm bill seemed likely to create the impression of farm groups against business groups. Since farmers were so often customers, this posed a serious marketing and sales threat that seemed more predictable and more costly than gains from a reformed farm bill.

The springtime publicity on the farm crisis chilled most of what remained of industry's ardor for open political conflict. The time appeared wrong. Legislators would be unresponsive. Business would be portrayed as kicking farmers when they were down. In a sense, the first farm suicides meant the last public words on business profitability.

That did not mean that business was ineffective, disinterested, and not part of the 1985 Food Security Act process. Organizations such as the International Association of Ice Cream Manufacturers, the Food Marketing Institute, and even the small National Independent Dairy-Foods Association kept after dairy, sugar, and peanut programs even during conference proceedings. Such user firms as Pizza Hut and Mars had active representatives, too. The Fertilizer Institute lobbied hard through the process. In a less obvious manner, the Grocery Manufacturers of America kept affirming their desire for a market-oriented bill.

The commitment of these groups added to the momentum in support of lowering guaranteed loan rates, the major contribution initiated at the urging of the academics. Along with several other organizations, activist business interests were equally instrumental in mobilizing lobbyists for key votes to defeat mandatory production controls. From a market-oriented perspective, however, those were the only victories for agribusiness in 1985, and both were gained with most of the Washington-based farm groups eventually in at least tacit policy agreement.

In retrospect, none of this is surprising. Several myths accompanied the belief that industry and business leadership could reshape the 1985 farm bill. When but a few fallacies became evident, the strategies of business-inspired change proved faulty, just as many veteran farm lobbyists with a historical perspective had suggested. Among the most important were the twin myths of business unity and shared self-interest. As noted in Chapter 6, many active organizations are not free traders. Archer-Daniels-Midland, for example, seeing the high levels of support in the sugar program as advantageous to its high-fructose sweeteners, was instrumental in gaining support from a few business lobbies while causing still others to remain silent. Its efforts did much to neutralize the impact of the sugar users' coalition and, indirectly, the opponents of dairy programs. Some firms even argued to move further away from unrestrained trade. For instance, ConAgra offered a plan for export subsidies.

Even more business unity gave way to the enthusiasm of both firms and trade associations for single provisions of unique concern to individual interests. Grain traders spent more time on grain quality than on grain pricing. Food processors saw clear producer title to commodities as the greatest need. The American Bakers Association wanted and got a wheat reserve. The open-ended structure of the farm bill gave business representatives, like commodity groups, the nearly unrestrained ability to pursue amendments of narrow and limited intent. As a result, some potential free traders had their lobbying resources tied up elsewhere and became very supportive of the bill's eventual passage in order to protect their gains.

Instead of finding a collectively well-financed and skilled lobby in the agribusiness community, coalition leaders encountered more a myth of rather than a potential for immediate power. It was a surprise to many trade association and agribusiness leaders that even the best-financed staffs had to limit their time and attention to selected priorities. In other instances, it was even more surprising to those individuals that many large firms and industries had staffs severely limited by small size, low pay, restricted budgets, and inexperience. Coalition meetings, rather than being the places of unbridled enthusiasm for policy change that some had hoped for, were plagued by participant inattention, lack of follow-through on assignments, leaks about strategy and tactics, and even a few spies from the opposition. In short, they were surprised by the implications of an agribusiness lobby composed mostly of single-issue and single-project interests.

The most damaging myth, however, was that all the provisions in the 1985 farm bill were equally open. In theory, but not in practice, anyone can get the ear of Congress. Agribusiness encountered the widespread congressional opinion that this remained a farmer's bill and that producer groups held the greatest legitimacy. After the farm crisis came to the forefront and dominated the bill's proceedings this opinion solidified, but as noted earlier, it was decades old. This situation had important consequences. Equally well-prepared information did not hold equal weight. Access to policymakers was often a problem. It also led to the need for agribusiness representatives to seek out even small producer groups to use as coalition partners on major provisions to front for the alliance. Consultants with long-standing agricultural policy ties were brought in and out of the process for similar reasons. Without such cooperation, agribusiness may have had some voice in farm bill proceedings but it would have been less consequential.

OTHER PARTS OF THE PUZZLE

After years of searching for a broader constituency for the farm bill, agricultural policymakers developed and passed what with one major exception could have been called the Farm Product and Export Act of 1985.[20] Yet that exception provided the appearance of broadly important social policy that policymakers seek in a farm bill; helped neutralize potential ex/al opposition to costly support legislation; and because of its resource protection provisions, assisted in the titling of what came to be called the Food Security Act.[21] In fact, that exception, with its soil and wetlands emphasis, initially was so popular as an integrating mechanism that many policy participants sought to give both the farm bill and the simultaneously considered farm credit bill some sort of conservation titles.

James T. Bonnen has argued that commodity interests gained their greatest importance when three changes were made in agricultural policymaking: farm policy undertook a new consumer orientation, commodity expansion rather than balanced production prevailed philosophically in writing a farm bill, and the USDA lost its ability to exercise leadership in initiating specific policy directions.[22] As interests concerned with single titles and provisions, commodity groups simply negotiated a series of individual settlements that finally became a bill. In 1985, both commodity groups and a number of other intensively directed interests benefited by similar conditions. First, the USDA lacked leadership on the bill. Second, nearly the entire Washington establishment of agricultural interests, in a rare display of action and unity, assumed a highly defensive posture against production controls as demanded by the agrarian reformers. Neither policymakers nor private interests could be dissuaded from making a concerted attempt to regain control of the international export market. As a consequence, the commodity experts—including those affiliated with both farm and trade interests—became the leading proponents of what became law.

Having cast their lot rather weakly with other user and retail organizations in an ill-fated opposition to dairy, sugar, and peanut programs, consumer interests were of negligible importance in the farm bill process. In 1985 these groups were unnecessary anyway since there were so many other advocates of consumer-preferred high rates of production and attendant lower prices. With the export enhancement position so attractive to farm and agribusiness interests, consumer advocacy would have been redundant support.

Consumer groups were also unnecessary to the farm bill from another perspective—their status as major representatives of an agricultural policy constituency having important social policy concerns. The conservation and environmental groups had moved in to replace them, not necessarily with that purpose in mind but certainly in producing that effect. After these ex/al interests successfully stamped the farm bill with their demands, the imprimatur of the ex/al lobby and socially relevant policy was at least superficially on the legislation, and that stamp of approval was on the bill with no appreciable commodity production losses or interference in the provisions or titles of importance to other interests.[23]

Conservation and environmental interests, actively involved from the outset with key legislators and USDA experts, had prepared to claim their provisions in the farm bill quite early as well. Both the American Farmland Trust and the National Audubon Society formed agricultural policy units in 1981. Serious policy discussions over conservation cross-compliance and a conservation reserve began about the same time. Cross-compliance, with its requirement of farmer participation in soil conservation programs in return for eligibility for commodity price support programs, was a prime concern of a number of participants having a distinctly narrow involvement with agricultural policy. Some USDA soil officials, several environmental and conservation interests, a number of academics who specialized in soil programs, and a few consultants believed that a carefully defined strategy aimed toward specific politically realistic goals could set some important precedents for agricultural policy. Once legitimized as part of the farm bill litany of titles, conservation programs having an environmental protection focus could gain attention in the future, like food stamps, Food for Peace, and other programs championed by nonfarm and nonbusiness interests. These potential participants spoke openly of joining such groups as the Food Research and Action Center and Bread for the World as expected players and, as a result, policymaking partners in future farm bills because of their probable emergence as private interest spokespersons for these new issues.

Between the 1981 and 1985 farm bills, several organizational occurrences gave impetus to these new ex/al goals. The commonly supportive ex/al lobby, as an informal coalition, fell in line with activist leaders who kept promoting the issue. Because of its farm-specific orientation, the American Farmland Trust became the de facto leader on the issue. Technical analysis, provided through foundation support, became the major instrument in articulating a moderate and intentionally reasonable message about soil conservation programs that not even input agribusinesses could become too angry about.[24] Finally, policy entrepreneurs provided critical assistance in promoting the issue, developing specific goals, advertising these alternatives, and maintaining the link between mutually supportive interests. The two most instrumental entrepreneurs were Charles Benbrook of the agricultural board of the National Research Council of the National Academy of Sciences and Ken Cook, a consultant on conservation issues who wrote widely on the topic and was employed by several participating organizations.

The impact of these events led to two important consequences. First, the better-funded sister organizations within the ex/al conservation lobby began to allocate large sums of money and other resource assistance to soil problems and issues. While several organizations mobilized constituents for congressional contact, the culminating public relations event was the decision by the National Wildlife Federation to dedicate its 1985 National Wildlife Week to a soil conservation theme. Second, the moderate and farm-directed stance taken by American Farmland Trust's leadership led to considerable public support from farm organizations. The NFU and the NFO were early coalition partners, and the Farm Bureau cautiously endorsed some of the developing proposals. While this support did not take the form of much active lobbying, farm lobbyists let it be known that conservation proposals would be useful as cost-cutting mechanisms that dealt with the problems of surplus in ways compatible with the intent of commodity programs.

As a result, four goals gained priority among ex/al participants and three of them, as separate provisions, were inserted into the farm bill with minimal conflict and surprisingly limited congressional attention. The first was the conservation reserve to retire highly erodible or environmentally threatened land that had been promoted as an attainable priority by American Farmland Trust and soil conservation experts since early in the 1980s. However, congressional offices assumed leadership on two provisions advanced earlier by other private interests and then supported by the conservation/farm coalition, the sodbuster and swampbuster provisions that disqualified farmers from commodity price support programs if they converted environmentally fragile erodible acres or wetlands to crop production after 1985. The fourth goal, conservation compliance, was finally written into other provisions by disqualifying those who farmed highly erodible lands without following approved Soil Conservation Service plans. The agricultural soil experts involved in the conservation coalition were both the initiators and proponents of this requirement, following it through House Agriculture Committee defeat, floor inclusion, and conference committee adoption.

While this victory was satisfying to the ex/al lobby, it was hailed in Congress and the administration as a major farm bill innovation as well.[25] Perhaps policymakers were most pleased by the lack of hostility that the provisions engendered while they conferred some degree of a sorely lacking reform status on the Food Security Act. Or, it may have been that policymakers were happiest with the obviously parallel fact that conservation provisions were not seen to interfere with the basic income support and export enhancement direction of the 1985 act. No one denigrated the exceptional lobbying tactics of ex/al lobbyists as they, for once, carefully worked with the established institutions of agricultural interests and policymakers. However, respondents uniformally expressed the view that conservation provisions succeeded only because they were largely distinct, autonomous, and politically useful in adding a cost reduction emphasis to the bill. These provisions were aided by the same contextual factors, both economical and political, that benefited commodity programs.[26] In short, conservation issues provided an early and agreeable national policy emphasis that many groups could rally around in finding reasons of both substance and coalition building for supporting the 1985 farm bill.

Less recognized measures brought the same kind of support to the final bill. Animal rights activists, largely through personal contacts with Senator Dole and House Majority Whip Tom Foley, were responsible for provisions to improve the treatment of laboratory animals. Food stamp advocates, with less government attention and support going their way, were devoting shrinking organizational resources to protecting programs already financially reduced in 1981. Bread for the World was similarly involved in Food for Peace provisions, as was much of the staff of Interfaith Action for Economic Justice. These organizations, which represented most of the ex/al emphasis on agricultural policy, directed their planning and lobbying toward provisions no less narrow than the commodity groups or any trade-oriented agribusiness association. As the USDA's Young Executive Committee might have anticipated when they made their suggestions in 1976: there was little time for involvement in broad issues of policy reform, there were almost no incentives for such involvement, and there were plenty of reasons for avoiding the conflict that such activism would have created with established interests. The organizations were best served by being advocates for provisions that balanced the eventual passage of the Food Security Act, a bill that, like its immediate predecessors, was a mosaic of individual parts and now varying policy directions as well.

IMPORTANCE OF FARM BILLS

As was the case with the data presented in Chapter 10, no organized interests dominated the proceedings of the 1985 farm bill. Both legislators and lobbyists, with USDA advice, occupied center stage, but not for the same purposes. It was left to Congress to put the final package together. Commodity organizations participated more actively than did any other private interests in negotiating final settlements, however.[27] During concluding conference committee markdowns, almost no other interests gained access, but major commodity groups were frequently called in for consultation, even assisting in drafting final provisions days after Congress had passed the bill.

Influence over the final outcome was shared, however. By identifying a farm crisis, the farm protest groups had given agriculture the collective presence to command attention in Congress and before the public. This capacity to demonstrate agriculture's presence meant that, no matter what difficulties arose, it was in almost no one's self-interest to prevent passage of the farm bill. This, along with the large margins on the final votes, indicated that it was unnecessary for legislative leaders to construct a carefully monitored winning majority coalition in Congress. When the details could be arranged and an allowable budget figure negotiated, a bill would pass. This became evident in July 1985, when a key budget resolution was agreed to by the Senate Budget Committee and the White House that only slightly reduced anticipated farm program costs.[28] Congress then ratified that resolution by clear margins, thus signaling an end to conjecture about the termination of commodity price support programs. The maintenance of high target prices, with the obvious prospects of escalating deficiency payments, could not have been achieved by commodity organizations without the groundwork and sympathy established over several years by agrarian reformers.

Another collection of private interest groups created the climate necessary for reducing and changing loan programs, the one commodity feature of the bill cited immediately after passage as an important departure in agricultural policy. Without the export enhancement and market-oriented concerns of academic agricultural experts followed by publicly aroused agribusiness interests, and without the tacit endorsement of consumer interests as well, commodity groups would not have agreed to what they considered "give backs" on loan programs. Yet by mid-1984, most commodity lobbyists knew that other interests—not to mention trade conditions—would be able to force worse policy losses if their groups did not give in on loan programs, in an attempt to lower U.S. commodity prices. Technical analysis and the need for an improved business export climate, as these were articulated forcefully by so many, influenced not only legislators but commodity group officials as well.

These sets of interests, working distinctively apart from one another and not toward any cumulative package, contributed substantially to the final outcome of commodity and trade programs in the 1985 act. None could have created these results on their own. Without the farm protest movement, deficiency payments would have been reduced and farm programs dismantled. Had the commodity groups been lacking in expertise and access, however, the protest efforts would not have produced a negotiated settlement that selectively assigned massive increases in federal expenditures to individual farmers by commodity.

Shared influence of the farm bill was no less evident in other provisions.

Conservation interests made major headway in gaining agricultural policy legitimacy by getting their own provisions into what became, for all practical purposes, their part of the bill. Other ex/al interests, the commodity groups, many agribusiness interests, and research interests all did the same thing, laying claim to selective policy wants and then seeking to procure them. Unlike the complex settlement in which target price policy went in one market direction while loan provisions went in another, these provisions were won and lost by groups and firms that had worked out what amounted to, in essence, jurisdictional turf agreements.

One major lesson emerges from the highly contextual resolution of the Food Security Act in 1985. The fragmentation of decision-making power within such narrow institutional corridors as congressional subcommittees benefits interest groups as organizations that, for the most part, are attempting to influence only small portions of agricultural policy. Interest groups with such narrow demands profit still more by decentralized congressional processes that make it feasible, especially in the Senate, for individual legislators to shepherd specific provisions of a major bill through the policy process without comprehensive oversight. When control over public policy is widely dispersed, every interest's potential for influence increases if its representatives can develop a successful strategy for translating public-policy goals into specific demands that make few enemies.

When this occurs, as both the structure of the farm bill and its institutional governing apparatus allow, policy participants will do whatever is feasible to promote narrow goals to the exclusion of comprehensive policy or even efforts to reconcile future problems resulting from provisions with inconsistent goals. During the 1985 farm bill negotiations, lobbying tactics of many types of interests resulted in several efforts to further undermine authority from the White House, USDA, and congressional leaders. Strategies were also worked through to discredit the agrarian protest organizations and move away from active involvement with general-purpose farm organizations, both of which exhibited diffuse rather than narrow policy goals.

Efforts to focus on selective provisions also led to strategies that purposely overlooked contextual conditions that interest representatives and members of Congress could not easily incorporate into their policy goals. Farm debt problems, as most participants acknowledged, were ignored as policymakers sought to handle the farm crisis through commodity programs. Meanwhile, the farm credit bill sought to aid Farm Credit System lenders rather than borrowers. On nearly every provision, the probable impact of the Gramm-Rudman-Hollings Act was considered and then put aside even though its results would be felt within a few weeks of the passage of the farm bill. This led to an immediate revision of the bill when Congress reconvened in 1986.

Other actions, again duly considered, resulted in other early backlashes against the Food Security Act. Minor commodity and other interests affected by planting provisions on diverted acres descended on congressional delegations when they understood the pending disequilibrium to be created by increased 1986 production of their crops by additional growers. This opposition led to a second 1986 farm bill resolution, or Farm Bill III, in midwinter. By fall 1986, another provision had been generating unfavorable publicity and considerable interest group complaints that led to yet another legislative change in the 1985 act. In order to allow increased deficiency payments when loans rates were lowered, ceilings on the amount of federal subsidies a farmer could receive were removed. Although USDA officials and several representatives of private interests had warned about the negative public relations consequences of such action, legislators were unconcerned until a California cotton farm received $20 million in program payments.

These problems produced a considerable amount of criticism of the 1985 farm bill long before its provisions had any opportunity to produce the results for which they had been intended. To those who had called for reform, it was a bad bill, but it seems likely that *any* farm bill—subject to so many highly selective demands and negotiated in so many subcommittees—will be inadequate. The high political salience of this complex and technical legislation created a situation in which many could win, but not what they may really have wanted.[29] Many otherwise satisfied respondents were not certain that their gains were positive for an American agriculture upon which their organizations were dependent. A number of private-interest representatives felt that their refined professional skills had not obtained what either their clients or agriculture really needed. Thus, the farm bill process enhanced even the self-image of lobbyists as partners to a wasted, yet necessary, profession.

Even the perceived importance of broadening the focus of agricultural policymaking to address some nebulously defined national policy problems and the desires of a broader range of private interests has decreased the prospects for a farm bill that does more than incrementally shift policy on a provision-by-provision basis. The farm bill process—divorced as it is from comprehensive agricultural policymaking and any integrated attempt to link farm, food, fiber, and trade problems—cannot be made easily responsive to an American agriculture that is changing in response to an interdependent set of worldwide problems. The influence of private interests—as played out in securing the specific provisions that policymakers then use to assemble the legislative coalition necessary to pass the bill—is only part of that problem.

12

On Influencing Public Policy

Based on an extensive array of responses pertinent to public policymaking in a distinct but complex issue area, the findings of this study deserve to be placed in a theoretical context. They are important because they advance understanding of the policy process. There have been few scholarly opportunities to generalize about the subjects of private interests, public policy, or American agriculture. Some previous impressions of those subjects must be confirmed, and a few others must be challenged.

In keeping with the audiences for whom this book was intended, two bodies of theory will be addressed: agricultural policy as developed mainly by agricultural specialists and interest group theory as advanced primarily by political scientists. Agricultural policymaking is addressed first because it is important to later concerns about organized interests. Interest group theory, or more precisely, interest theory, is discussed in the next two sections. A few observations are made about the developing policy-relevant literature, with particular emphasis on the research that followed the innovative thought of Mancur Olson. Because of the specific emphasis of this book, the last section examines the more recent theoretical concern with the meaning and implications of an expanding universe of organized interests in government.

AGRICULTURAL POLICYMAKING

As a twentieth-century replacement for generalized social reform legislation, agricultural policy created a series of economic supports for a farm society.[1] These new supports originally emphasized, first, the regulation of farm markets, input, and products; and, second, the creation of a physical infrastructure of credit, roads, electricity, and other investments intended to bring direct benefits to farm production.[2] The latter, for the most part, formed one portion of a 1930s explosion of farm economic support programs that also set in motion what became the dominant emphasis of agricultural policy. Because of the commodity emphasis of these programs, farm policy began its long romance with programmatic decisions that emphasized the selective assignment of public policy benefits to farmers.[3]

The selective assignment of benefits to farmers, with its Depression Era climax, set in motion the beginnings of an incrementally evolving agricultural policy. Price support legislation, farm credit, soil conservation, and marketing order agreements became the now five-decade-old centerpiece of federal farm policy intervention and supply management.[4] This initial emphasis on selective benefits should not have been surprising; farmers had been identified by a wide spectrum of observers as special in their social, political, and economic importance. The financial climate of the 1930s, with its socially threatened small-farm life-styles and its mobilization of political representation on behalf of numerous specific constituents nationally, made it easy to couch the problems of agriculture in economic terms. Eventually, agricultural policy meant farm-specific economic policy.

The continued use of economic policy, in which benefits were assumed to be allocated quite logically on an increasingly selective basis, had an established analytic justification: under the norms of surplus conditions, farmers cannot gain fair returns for their labors and investments without a regulated marketplace. So, surplus had to be managed to sustain the incomes of farmers. Beyond that, marketplace intervention was judged necessary because the biology and associated uncertainty of commodity production necessarily leads to uncontrolled supply fluctuations that cannot be adjusted quickly under short-term conditions. Moreover, product perishability leads to even more regulatory needs; interruptions in the distribution of commodities will create an irreplaceable loss.[5] The presumed need to control for these conditions mandated an extensive, not to mention expensive and thus beneficially attractive, interventionist role for national government in an effort to maintain both farm incomes and a long-term production capacity for the future.

Alden C. Manchester outlines the concerns that can prompt government's policy intervention for a single commodity, in this case dairy. Manchester enumerates the situations that can develop within the interventionist framework to provide incentives for private interests to attempt to influence policy decisions on pricing matters alone. First, a large number of fluid milk producers are at the potential mercy of a comparatively small number of dairy processors. Second, once legalized, long-term contract commitments can be advantageous economically to both processors and producers. Third, terms of exchange between the two parties require regulatory action to provide security, equity, and longer-term responses to changing milk market conditions. Fourth, if legally possible, producer groups can negotiate more favorable contract terms with processors than can single producers. Fifth, with a public agency as arbitrator, these contract relationships can be negotiated to ensure greater stability of supply. Sixth, some institution with sufficient expertise about fluid milk, if legally permitted, can be used to limit still further problems of market instability by protecting supplies, costs, prices, and delivery.[6]

Beyond this conceptual justification of price-supported agricultural policy, severe disagreements exist about how balanced policy decisions must be between farm income supports and market competition. Since price adjustment programs went into effect, policy analysts have divided on the question of controlling production. Should surplus be reduced by production cutbacks or market expansion? Relevant side issues, such as the resulting effect on land prices and eventual costs of commodity production, have produced similar points of unresolved policy disagreement as government officials keep returning to problems of surplus management in which reduction issues dominate all long-run problems.

On balance, despite historical cleavages among agricultural specialists, economic analysts have favored policies that, depending on farm needs and the economic context in which issues are decided, encourage commodity prices close to what would be established otherwise as market prices. As surpluses grew even larger in the late 1970s, accompanied by unexpected declines in foreign sales, this market-oriented approach gained greater urgency for the experts.[7]

Policymakers did not respond wholeheartedly, however, so agricultural experts began again to look at the agricultural policy process to understand why not. In the 1950s, it had been clear that general farm organizations differed in the value their representatives attached to farm policy. The high-support orientation of the Farmers Union with its message that "prices are made in Washington" was greeted vociferously by Farm Bureau leaders who looked beyond domestic to international markets.[8] In essence, leading organized farm interests stood as competing spokespersons for the two analytical positions influencing the thought and sometimes separating agricultural experts. Moreover, these farm interests articulated their concerns by arguing for a comprehensive unity of national purpose in setting commodity policy.

By the 1970s, however, the policymaking environment had changed, although the overall causal dynamics of that change and its associated variables remain open to different interpretations. Agricultural experts noted an even greater concentration of production in the largest farms and an accompanying increased specialization of farms by product. For activist farmers, there were more incentives to assign policy benefits even more selectively by commodity. The decentralization of power in congressional committees and the associated decline of presidential and partisan leadership aided the commodity interests. These interests had long been organized around commodity programs but had only in recent years been able to exercise policy leadership. Farm policy experts from price support proponent Willard Cochrane to market-oriented Don Paarlberg recognized the phenomenon of fragmented policy articulation and resolution and agreed that the situation made it impossible to address the common problems of agriculture in any consistent or stable fashion.[9] Because of the convergence of a changing agriculture and its private-interest representatives, and because of a decentralized policy process, agricultural policymaking was now beyond the pale of major reform. As single-product supply problems shifted and political fires were fanned, it was left to be adjusted continuously by those who made the rules.[10] Adding to the instability of policy decisions was the proliferation of still other interests, the "new-agenda"-directed ex/al organizations that made unfamiliar social-policy-directed demands that meant still further adjustments.[11] Agricultural policy was being made more frequently in unfamiliar policy institutions. There appears to be considerable accuracy in such a theory of policymaking, but cause and effect seem to be confused, and the scapegoats of the theory are private interests, especially commodity organizations.

There is no evidence, however, that these organized interests have ever had the influence, individually or collectively, to set the policy agenda within agriculture or to adjust it. The blame seems to fall on the policy process, of which interest groups are just a part. Any such blame—or for that matter, any virtue attached to the ability to make piecemeal adjustments without changing the policy whole—should be assigned to a factor long addressed by the traditions of political science, the structure by which policy decisions are made. When policymakers of the 1930s decided to make farm economic policy by assigning benefits to distinct commodity producers, they unwittingly set into motion a rewards process that policymakers later would adjust to for better monitoring. Private interests would also react and adjust to the process to get what they could from government. In that sense, government officials seem to have treated agricultural policy no differently than they did other policy areas: by emphasizing specialization and division of labor in managing product supplies.

It must be remembered that in the 1920s and especially in the 1930s both government and industry managers developed policies on the principles of Taylorism and other aspects of scientific management.[12] As practitioners of a cult of efficiency, the policymakers who addressed problems—including those of agriculture—for the Roosevelt presidency favored professionally administered responsibility and accountability, as in the USDA and the secretary of agriculture; role differentiation, as by commodity; and performance standards, as in the ability to monitor policy success in terms of specific measures of price and crop surplus.[13] The already existing degree of commodity specialization in the agricultural sector fitted well with such predilections in the Roosevelt administration and its Bureau of Agricultural Economics. When the commodity groups transcended the general farm organizations in policy importance, they did so at government's unintentional request. These previously regionally influential organizations gradually gained more national influence as decision rules evolved in their favor. Commodity interests in agricultural policy only set into motion a continuing intensification of narrowing demands and government responses.

SOME CONCERNS ABOUT GROUP THEORY

From the outset, this analysis has rejected the idea that relevant private interests as active policy participants include only interest groups. Traditional interest groups are joined by other nonparty organizations, each very different in form but all nonetheless organized to influence public policy decisions. As Chapter 11 makes clear, all of these organized interests matter in that they have some impact on agricultural policy. They matter because of the complexity of those variables that enter into any policy decision.

Policy decisions are influenced as organized private interests either directly manipulate or assist policymakers in handling three sets of variables: the formal and informal rules by which policy is made, positive knowledge or the analysis of policy problems and their solutions, and value knowledge about the emotional and traditional appeal of certain policy proposals and goals. As seen throughout this book, different private interests are able to capitalize on their control of different portions of different variables. Although, for example, commodity groups have benefited from decisions on farm programs, agrarian reformers have captured the imagery of family farm values and the place of those values in American lore. Other interests matter because they have found a productive policy niche for their representatives.

It must be emphasized that these policy niches are determined by both what private interests offer to policymakers and the degree to which policymakers accept as useful what interest representatives can provide. From the research in this book, the latter seems the more important. Goals are determined by the acceptability of policy demands and by the likelihood of victory.

Any study that hopes to assess public policy impact, therefore, must consider several forms of organized private participants other than interest groups that share input through diverse representational styles and strategies—reform movement organizations, single firms, and the cadre of institutionally affiliated experts. All of these groups are advocates of important policy alternatives for agriculture and repositories for distinctive policy ideas that interest groups either will not or cannot articulate because of their own organizational problems. Moreover, all of the different organizational forms, dependent on the degree of intensive or diffuse policy focus, provide unique resources that are applied differently in order to create the appeal that will maximize an opportunity to be heard and heeded. Protest and challenge, as practiced by the reformers, and organized public policy education, as propounded by agricultural experts, are effective in gaining policy attention and, therefore, getting involved with policy decisions. However, a multipurpose business firm in many ways behaves more like a similarly structured trade association than like another business that assumes a single-project policy identity. In brief, determining who represents what as a relevant private interest is impossible if only *groups* are considered.

Two factors seem especially important for scholars in this expanding view of which organizations should be studied as private interests. First, any conceptualization of group theory that purports to address the policy impact of private interests in any noncomprehensive way is too restrictive. To make important advances in interest theory, the focus of attention must be broadened to encompass more than just formally organized associations. One example will suffice to make this point. The Corn Refiners Association (CRA), an established and venerable interest group, has a limited policy scope and a reluctance to send its representatives into certain policy disputes. Archer-Daniels-Midland, as a major corn refiner, independently mounts a far more aggressive lobbying strategy than other refiners favor or practice. Any attempt to understand the impact of corn refiners by focusing on CRA would grossly underestimate the industry's political impact since ADM's lobbying is not something that goes on within the context of intra-organizational CRA politics. Strategies to exercise policy influence have become that specialized by interest.

Second, if several organizational forms are to be considered in the further development of interest theory, lobbying must be discussed—but not necessarily redefined—as an encompassing political strategy that involves many tactics. It does little good to think of protest as something groups do because they are otherwise unable to gain the access that established interests possess. Both James Q. Wilson and Charles Tilly made this mistake.[14] As agricultural policymaking in the 1980s has demonstrated, agrarian protesters have served as vital and frequent policy participants with other strategic options. Other organized interests have depended on those activists for their own successes, but, most importantly, the agrarian protesters have chosen their lobbying strategy for a specific reason. They understand that their goals need to be tied to generalized reform demands and that such a diffuse focus will only get lost in the labyrinth of Washington interest group politics. For the most part, their representatives do not want to become an established Washington interest, even though on certain occasions they borrow from the conventional tactics of politics.

For similar strategic reasons, several agricultural experts, including both academics and consultant technicians, manipulate policymakers in ways that are hardly conventional. Their intended behavior, from using established access to presenting useful research in exercising policy influence, is different only stylistically from that of association staff. In their haste to label one political act lobbying and another mobilization, scholars have regressed to hairsplitting. They have also forgotten Bauer, de Sola Pool, and Dexter's warning that a lobbyist's job must be understood in terms of many recurring tasks, only one of which is *making* direct political contact.[15] An increasingly important one, through both coalitions and the use of constituents, is *to* direct political contact. The research presented here also makes evident something many policymakers want acknowledged: that there are many individuals whose jobs make them de facto lobbyists even though they would never want to register as such. Interest group theory, in this regard, presents academic problems because it disguises lobbyists and lobbying as well as the actual lobbies.

This criticism of the restrictiveness of interest group theory, as a body of knowledge whose practitioners want to study too little, only modestly advances Wilson's important work in *Political Organizations*.[16] From Wilson, students of interest group politics can see that contextual circumstances do matter in the way individuals organize to represent their ideas to government. Depending on the entrepreneurs, patrons, and members, organizational form may vary. Depending on what these individuals want, whom they want it from, and what ideas they represent, that form will vary even more. A great deal of tolerance must be allowed for the diverse structural characteristics assumed by those private interests that emerge as active lobbies. Even noncomprehensive, selective public policymaking will not encourage only the formation of increasingly intensive interests. It may also lead to a backlash of diffusely organized reformists that move more and more in an alternate direction.

Yet, in private interests and public policy involvement, there are still substantive bounds to the analytical units of analysis. A new theory of organized private interests would begin only by acknowledging that the basic units include associations that are not voluntary.[17] It would then go on to examine the aspects of organization that have been critical to the policy emergence of active private interests. This study helps organize some parts of that theory and provides some advances on previous findings. These findings and aspects of interest theory on which they are based can be reduced to several key points, some of which will need further elaboration in discussing the new universe of interests.

1. Intra-organizational Relationships. As political observers expanded on the seminal work of Mancur Olson, it became clear that organized interests were not monolithic institutions in which policy goals were neither shared nor shaped jointly by all. Olson noted that individuals within organizations are "primarily interested in their own welfare" rather than that of the unit of which they are a part.[18] Organized interests, in order to avoid free riders who benefit from an organization's political influence but who refuse to join and offer support, typically provide selectively assigned service benefits to members. An encompassing literature has shown that these services as well as an organization's involvement in policy decisions are responsible for keeping successful interest groups together. Public policy involvement by itself, however, seems insufficient to recruit and retain the body of members needed by organized interests to survive. Even those who do not accept a rational-choice approach agree to that.

For each of the types of organizations and for the different forms of organization in this study, keeping supporters involved in the political life of the organization is a problem. Agrarian reformers, representing the most comprehensive policy demands, need selective services to stabilize constituent commitment. Narrowly intensive single-project business firms, on this count, follow strategies of episodic policy involvement because their executives do not see the value of continued policy involvement. It is not just voluntary associations with Washington staffs that are forced to provide an immediate and personal relevance for those who pay the group's bills.

2. Moving toward Policy Involvement. Robert H. Salisbury took Olson's theory of individual self-interest and incorporated the importance of leadership.[19] Entrepreneurs establish organizations, structure selective services to get and keep members, and then exercise discreet organizational leadership over a group's public policy activity.

In keeping with entrepreneurial theory, the importance of leadership to diverse types of newly emerging agricultural interests has been essential. The farm protest groups were created by leaders who invested their organizing resources in these movement style lobbies even though their reputations were built on the contagion effects of widespread social discontent and economic hard times. The origins of the ex/al interests of the 1970s are tied to neopopulist activists whose project loyalties shifted from one new and often small organization to another. These leaders recognized the circumstances in which policy windows were likely to open for alternative agriculture, laboratory animal rights, and many other causes that needed to be sold to supportive patrons before further action could be taken.

Policy entrepreneurship, as identified among agricultural interests, has an importance that does more than just bring new groups toward active policy involvement, however. Most agricultural interests exhibited a policy activism that was associated closely with member services, nonpolicy association activities, business products, or even the business style of a firm's executives. The organization's staff typically reacted to policy events and circumstances that affected these heartland concerns. Unique policy activity from any single organization has not been the result of responses from watchdog lobbyists or the warnings of in-house financial analysts. Agribusiness firms and trade associations have been moved by innovative staff who, for example, promoted their own questions about cheese distribution programs and dependence on the dairy program or the implications of farm policy for such new business ventures as biogenetics. Farm organizations that make major changes in their policy positions are being responsive to entrepreneurial promotion as well. Both the movement of the Farm Bureau to promote a high-price-support version of the 1985 farm bill and the final willingness of the Farmers' Union to accept mandatory production controls resulted from internal entrepreneurship by factional leaders within the two organizations. There appears to be a marked tendency toward policy stagnation and repetition for private interests that lack entrepreneurial staff who are especially willing—and able—to react to changing policy and environmental conditions. Without such venturesome activists, interests do not appear to be vital policy forces in the midst of changing agricultural conditions.

3. Limiting Entrepreneurship. Entrepreneurship in developing new organizations has few restrictions. Newly organized interests emerge as policy activists as long as there are, in some combination, leaders who can sell a stated need and followers who buy it. Interests as organizations will proliferate because entrepreneurial leaders can range from innovative service providers to charismatic orators who create only the impression of likely benefits. Patrons and affiliates of new groups are not always rational or sophisticated consumers. The American Agriculture Movement created an ongoing interest group from a farm strike that not even its leaders supported. Agribusiness funded the reemergence of the Agricultural Research Institute in the vague hope that a supportive rather than obstructionist ex/al interest could be created and gain policy legitimacy. Most of the new ex/al groups acquired funds from individual and institutional contributors by playing to the ideals, sympathies, and fears in a climate in which specific issues were becoming topics of concern.

However, as the agrarian reform movement best demonstrates, entrepreneurial success in creating an organization ensures neither longevity nor specific policy success. Organized interests, whether they are business firms or grassroots local, must show some relevance to survive.[20] Securing benefits for selectively assigned service is one way; as, for example, when a Washington office for a large multipurpose company secures a favorable ruling on a new product from a slow-moving Food and Drug Administration, or when a state farm crisis committee intervenes with the Farmers' Home Administration in debt restructuring for a troubled farmer. In both of these very different instances, the representatives of the organized interest are creating exchanges that are favorable to its continued existence and that show specific reasons why the organization should have a public policy mission, even if that mission points to the need of intensive policy concerns for even a broadly reformist organization.

The examples above show how important internal maintenance and external goals are for agricultural interests.[21] Entrepreneurs have fewer options to pursue in operating an organization than they did in founding it. First, there are the complexities of the policy process; second, the unlikelihood of procedurally ignorant activists' gaining specific responses from public officials; and third, the need to show backers a reason to maintain their financial commitment. Consequently, private-interest representatives are restricted environmentally as they maintain necessary relations with both government officials and financial supporters and, frequently, other organized interests as well. Even if the organization is one in which representatives reject the usual strategies of cultivating and relying on direct access to public officials, they are still seeking relationships that elicit specific responses when the need for those responses appears again. Therefore, policy entrepreneurs become more conventional as they move from a situation in which they are organizing to one in which they seek to keep the interest together. These conventions, in fact, cause representatives to think and act as policy managers who deal in adjusting and proliferating narrow demands rather than as venturesome entrepreneurs who work to promote change. Perhaps this explains why the agrarian reform organizations have always been reorganizing, never able actually to develop and initiate their *own* proposals.

   4. Developing a Policy Identity. Because of these operational dynamics, the agricultural lobby—and presumably other sets of interests too—exists as a collection of primarily well-established organizations with some newcomers always seeking entry. In struggling to survive and retain the capacity to become involved efficaciously in public policy decisions, organized interests have each had to confront the question of which issues their representatives should address and under what conditions they should do so. While procedures for determining these priorities usually involve much of the organization, de facto leadership falls to a few policy activists who were hired for their skills or elected because of their organizational involvement. However, few organizations move far from their members or patrons in deciding their policy options, both responding to and cultivating their views on pending issues. Private-interest respondents to this study found no lack of concern among those who supported the organization financially. On the contrary, crisis conditions made for careful monitoring by at least some members or patrons. At the same time, few organized interests will decide to move far from the major concerns and identified problems that occupy relevant policymakers. Even conservation interests found 1985 an appropriate time to link their goals with commodity programs.

   The leadership is left with a limited calculus to reflect on a variety of wants: deciding whether to cultivate an access-dependent lobby, maintaining and not exhausting relationships, developing a reputation for influence as a skilled lobby, selecting winnable positions, and managing lobbying resources over time. The conventions that confront the policy caretakers of any organized interest are not likely to produce radical organizations that take risks, define their interests grandiosely, redefine their positions frequently, or move much beyond the incremental pace of agricultural policymaking. That is, most interests decide on a political identity that others can understand. Then they generally stick with that identity, developing positive feelings of success. Thus, at any one time, most organized interests will be conservative—but certainly not unaggressive—by nature. As a result, they are unlikely to be responsive to the problems of an internationally changing agriculture or a widely identified need for changes in agricultural policy that even their own representatives privately might agree is desirable.

   5. The Importance of Context. Changes always confront private interests, nonetheless. When the context of policymaking changes—as it must over time—some policy identities, which are based on policy components that become less valued by policymakers under the new conditions, will have less relevance and will bring those private interests that have them less opportunity for successful lobbying. Other interests will have greater policy relevance and will gain more opportunities to win policy rewards. Entrepreneurs and organizational supporters will create internal pressures to adjust identities marginally to best take advantage of the most significant policy variables. This altering of who represents what and who has the greatest influence also changes the dynamics of the interactions among a great many policy participants. It does so slowly, however.

6. *A System of Organized Policy Niches?* Conservatively maintained interests that seldom risk offending or confusing supporters make collective decisions about what their identities will be; and those decisions produce what might be understood as a system of interests within agriculture that parallels the components of the food system that was described in Chapter 2. Given the entrepreneurship that characterizes the formation of organized interests and the tendency to believe in the inherent value of organized political action, it is not surprising that the specific issues and problems confronting agriculture are blanketed by a growing number of active interests. That is, everyone in the food system seems to be organized. The irony, of course, is that this comprehensive set of interests does not address agricultural policy in either a comprehensive way or in some fashion that reflects the systemwide changes in food, fiber, and tobacco industries. Why?

The answer to that paradoxical problem derives from the fact that these interests do not form an ongoing set, or even sets, of organizations whose interactions jointly address points of common policy need. Coalitions of multiple interests are short-term enterprises, currently without much potential to mobilize and maintain the commitment of organizations that would rather protect their own private agendas. Most interest groups must avoid many issues on which outside observers expect them to be active because of both intraorganizational disagreements and the desire to sidestep disputes with members, patrons, policymakers, and other organized interests. Hugh Heclo, not surprisingly, described the composition of organized interests in issue networks as more like an amorphous cloud than an organized form.[22] The organizational circumstances of interest operations just do not make for stable issue networks or power clusters despite the still important concern for maintaining relationships with policymakers in distinct issue areas of agriculture. Too many organized interests, because of their episodic single-project or highly diffuse and overly extended policy focus, see relationships maintained by means other than the cultivation of continuous access on all potentially relevant policy matters. Others, better known and prominent in Washington, are marginally adjusting their goals to what are usually only subtle shifts in the agricultural, economic, and political context.

Organized interests simply find their substantive policy niche and usually fight to stay there as long as staying is feasible. Their greatest competition is heard among the babel of both posturing and active interests, not as a result of their engagement in conflict over policy matters. Most private interests are intensively focused organizations with such narrow concerns that they almost never address issues of general economic or social consequence. Many of these interests are too deeply immersed in the problems of commodity programs or specific ex/al conditions to share in generalized reform efforts. Even more of them, however, are organized to be only rarely involved in policy decisions on an even narrower basis.

Most multipurpose organizations are not much more capable of addressing a changing agriculture when confronted by the current decentralization of Congress, the lack of centralized policy leadership in the USDA, and the opportunity to address manageable policy decisions of narrow intent. Groups such as the National Cattlemen's Association have become multipurpose because of the complexities of the industry, whereas firms such as Cargill have done so because of strategies of business diversification. If anything, the diversity of internal subinterests within these groups restricts rather than facilitates active position taking on important issues. Other multipurpose groups such as the Farm Bureau and Food Marketing Institute have occupied more generalized policy niches. These interests, however, are too closely identified with ideological positions to be widely accepted by other interests and too narrowly directed to major industrywide concerns with commodity policy or sales promotion activity to serve legitimately as peak associations and become the aggregators of reform positions. Nor have any of these groups emerged or been accepted as effective brokers when competing interests disagree on policy positions. While consultants have been employed as brokers as a result of their own substantive and procedural niches, theirs is also a narrow rather than comprehensive policy role.

In moving organizational entrepreneurship into a more conventional concern with the restrictions of policy management, has the policy process created such a loosely structured system of organized interests that the interests play no part in identifying and focusing political attention on the critical issues of a transitional agriculture? There seems to be considerable validity in that observation, since organized interests have few incentives to foster changes that might well render them less relevant to policymakers. More attention needs to be given to this proliferation of agricultural interests in terms of the policy implications of what several political scientists have referred to as the new universe of interests.

MANAGING THE NEW UNIVERSE
OF ORGANIZED INTERESTS

There has been a great transformation in the representation of agricultural interests from that time, especially in the 1950s, when the farm lobby was split between two diffusely directed generalist organizations divided over competing directions for a comprehensive price policy. And that change is even more marked than it was in the 1930s when a few organizations could agree on policy reform issues and assist policymakers in developing new farm support programs.

That lobby has grown into one of a great many specialized interests, most of which are not regular policy players, organized around highly specific goals. It is tempting to conclude that this new universe of interests lends itself to, if it does not produce, an exceptionally unstable agricultural policy. Each time that changing conditions affect the issues and positions that an organization identifies as important, its representatives typically promote narrow corrective policy adjustments. In that sense, whatever causal effects an interest may produce are short-term, situational, and highly dependent on the context of the policy decision at hand. There can be no clear and lasting model of how influence is distributed that might best satisfy an economist. Agricultural economists are surely correct in their assessment that no policy positions are resolved for long, however. The new universe of agricultural interests understandably confuses policymaking by its diversity of cues and demands.

This does not mean that the impact of organized interests should be assessed negatively. As can be seen throughout this book, the agricultural lobby reflects a pragmatism that allows many contributions to better-informed public policymaking. Three such contributions, in particular, stand out. First, organized interests have been responsible for initiating and/or being the major proponents behind specific proposals that have brought distinct innovations in policy tools as well as important alterations in agricultural policy. Second, forceful efforts to represent specific clients and ideas have directed public and policymaker attention to social and economic problems that have national if not worldwide implications. Third, even less active organizations serve to keep many important policy relevant ideas alive and problems in view that might otherwise be forgotten in the haste of governance. These three contributions underscore the interdependence that exists between private interests and other policy players in making adjustments to a changing agriculture. Moreover, private interests are important as maintainers and protectors of agricultural programs.

Confusion and agricultural policy instability, and perhaps instability in the application of policy tools, are necessary by-products of the contemporary American policy process, nonetheless. Many policymakers, monitoring and responsible for many policy provisions, look to or at least listen to many organized interests in order to deal with what has become a troubled agriculture in the 1980s. There is none of the comprehensive policy leadership that many agricultural experts and even policymakers call for from that collective lobby; organized interests spurn peak associations, strong and disciplined multiple interest coalitions, and the recognition of certain types of organizations as the legitimate private interest spokespersons for general policy reform.

The system of private interest representation, despite its rather orderly transformation, is no less confused. The development of a commodity-based agricultural policy brought about the ascendancy of commodity interests as intensively focused proponents of selectively assigned farmer benefits. Commodity organizations with co-op connections had even greater advantages as they sought selective benefits that accrued to an even larger component of their portion of the food or fiber system. With the development of policy incentives that assigned similar rewards to agribusinesses and ex/al supporters, another set of narrow interests joined the agricultural policy process. Typically, these newly active interests followed a strategy of trying to bring decision making into a situation in which they could win and where established farm interests also could win, or at worst lose very little. In becoming conventional agricultural players, these new interest groups—despite their expressed intentions to the contrary—have preferred to back away from serious conflicts that may have pitted farm interests against those representing the "new agenda" or agribusiness. Skirmishes, such as those over sugar and dairy in 1985, have been minor and have not yet served to mobilize actively many interests on either side of the disagreement. Farm policy victories of the past, such as the passage of the food stamp program and the consumer revolt of the 1970s, depended less on direct conflict over policy goals than on the ability of ex/al interests to ally themselves with legislators who could make trades on their behalf.

It remains in this context that the old and once formidable generalist organizations still linger. They aspire to make many contributions: as brokers of conflict, as coalition leaders, as broadly representative policy reformers. All are necessary functions if organized interests are in any major way to address more than isolated problems and affect more than single, usually recurrent policy provisions. But, despite the need, interest groups have not been able to perform these functional services.

The failures of general farm organizations have resurrected yet another type of organization that seems out of place in dealing with a specialized and complex agricultural policy—the agrarian protest–style group. Were other interests capable of representing the problems of debt-plagued farmers, these groups would never have found a supportive constituency. Because of their place in the conventional community of Washington interests, however, other farm interests were compelled to act more moderately than were the protest movement leaders.

The resulting proliferation of groups, both in number and type, has brought a multifaceted lobby with widely varying points of view to agricultural policymaking. Because critical views—if not active lobbying efforts—confront one another regularly, a form of pluralistic representation that has not been present since the 1920s has been restored to agricultural policymaking. However, it seems an odd mixture of public posturing, diverse tactics, and interests of unequal standing on the heartland issues of agriculture. In addition, it seems questionable whether many of the newer nonfarm interests really represent the values or concerns of any significant part of the American public.[23]

The net effect produces a universe of agricultural interests that divide up the spoils of public policy among many beneficiaries. but the overwhelming emphasis, with the high visibility of farm protesters and the unique compatibility between farm organizations and commodity programs, produces a strong favoritism for farmers as the prevailing and highly selective service center for production-oriented agricultural policy.

In that sense, little has changed even with the disintegration of the classic governing subsystem in which farm interests and a few policymakers once independently directed funding and programs for agriculture.[24] The new universe of interests has produced one more condition with which organized interests have learned to cope in the pursuit of policy goals. While the system of farm representation is so shaky that its future is suspect if opposition mounts, lobbying tactics have been enhanced through the greater use of constituent pressures, mass mobilization techniques, PAC funds, and consultants. In that sense, farm interests protect themselves with more of the same strategies that characterize other American interest groups.[25] They have also made use of many of the policy positions taken by institutional representatives of business, universities, and foundations that have become more a part of the entire interest group universe.[26] Not all of these representatives have been opponents of farm programs. Adjustment to the patronage of ex/al interests has been tense, but the necessary accommodations that promote coexistence with these interests appear to have been made as well.[27] In short, none of the features that expanded the universe of organized interests in agricultural policy derailed the allocation of substantial benefits to farmers. Farmers are troubled, in the main, by their own representatives' collective inability to resolve their differences about the direction agricultural policy should take.

Will the inability of farm interests to lend stability to the policy process prove fatal? That seems likely to depend on how costly commodity programs become in an era of budgetary concern and whether agribusiness and ex/al interests lose their reluctance to oppose farm subsidies. An inability to come to a consensus on policy approaches has not been dependent on farm interests. Liberal and conservative farm organizations never resolved their differences in the 1940s, 1950s, and 1960s as policy tools were altered by the passage of administrations. In that sense, policymakers have always been confused about the proper way of directing agricultural policy and managing its surpluses. They did make decisions, however, and specific policy tools were placed into action.[28] Understanding how they have continued to do so is another research project that looks in greater detail at policymakers' responses, but for the foreseeable future it seems safe to say that there will always be farm bills that finally pass, even though that piecemeal format might well be the major impediment to actual agricultural policy reform.

Although the confusion of interest demands are greater, certain factors may minimize this perplexing impact. The analytical capacity of government has expanded. Moreover, legislators have developed a greater capacity to understand and respond to public opinion, relying less on what organized interests claim certain types of citizens want. While it may have been more accidental than intentional, the Food Security Act of 1985 was finally shaped more in accord with pub-lic-opinion-poll preferences of farmers than it was with what any organized inter-est or collection of organizations set forth or brokered as policy demands.[29] Inconsistencies among policy goals and all, farmers got much of what they wanted. Policymakers, with their attention turned to the electorate, once again refuted Earl Latham's claim that American government, even as an interest group society, is a pressure-group-driven model.[30] Rather, in that such a response took place in the context of such conflictual posturing between many publicly an-tagonistic interests, it recalled E. E. Schattschneider's warning that when a fight breaks out the crowd plays the decisive role.[31]

It must be concluded, as it probably should be in a democratic society, that the new universe of interest groups is beyond manageable proportions. Depend-ing on one's own political values, and the way in which those values relate to a concern for comprehensive agricultural policy reform, James Madison's fear of mischievous factions can be either laid aside or raised anew. It appears beyond the capacity of any single interest to dominate policymaking in a single issue area as it can do in the corporatist states of Europe.[32] The complexities are too great and the governing responsibilities too widely scattered among many centers of pol-icymaking for that to happen in the United States.

Because agricultural policy with its commodity orientation seems to be far more fragmented and narrowly based than other issue areas, perhaps this claim cannot be made for all.[33] Michael T. Hayes's statement that there are many policy processes in American government is true.[34] Organized interests behave some-what differently in different contexts, but probably not markedly. Regardless of type of policy, interest groups are subject to the same expanding universe of in-terests, conventional standards of cooperation, and need for a relevant policy identity as they are on issues of agriculture. As a result, the behavior of the many agricultural interests examined throughout this book seems likely to be repre-sentative of others who represent divergent views in American public policy.

Appendix A
Respondent Organizations
by Category

I. AGRIBUSINESS AND RELATED ASSOCIATIONS
   Agribusiness Council
   Agriculture Council of America
   American Frozen Food Institute (and Frozen Potato Products
      Institute, International Frozen Food Institute, and National Frozen Pizza
      Institute)
   American Petroleum Institute
   American Seed Trade Association
   American Society of Agricultural Consultants
   Animal Health Institute
   Chamber of Commerce of the USA, Food and Agriculture Policy Unit
   Fertilizer Institute
   Food Marketing Institute
   Futures Industries Association
   Grocery Manufacturers of America
   Irrigation Association
   National Agricultural Chemicals Association
   National Council of Agricultural Employers
   National Council of Plant Breeders
   National Food Processors
   National Frozen Food Association (Giuffrida Associates)
   National Stone Association

II. AGRIBUSINESS AND FARM CONSULTANTS/REPRESENTATIVES
   Abel, Daft and Earley
   American Agribusiness Associates
   Consultants International
   Economic Perspectives, Inc.
   Food Systems Associates

Gray and Company
Hauck and Associates
Heron, Burchette, Ruckert and Rothwell
Jaenke and Associates
Lyng and Lesher
Patton, Boggs and Blow
Quinn, Arthur Lee (Law Offices)
Schnittker Associates
Shaw, Pittman, Potts and Trowbridge (Farmers for Fairness)
Sorkin, Martin (private consultant)

III. AGRIBUSINESS AND RELATED CORPORATIONS
Archer-Daniels-Midland
Bunge Corporation
Cargill
Dow Chemical
Electronic Data Systems
Ford Tractor
Marathon Oil
Monsanto
Ralston Purina
(Three other respondent firms requested anonymity)

IV. COMMODITY ORGANIZATIONS AND TRADE ASSOCIATIONS

A. COTTON, TOBACCO AND PEANUTS
American Peanut Products Manufacturers, Inc.
National Cotton Council of America
National Peanut Council
National Peanut Growers Group (White and Associates)
Plains Cotton Growers
Southwestern Peanut Growers Association (Meyers and Associates)
Tobacco Associates

B. DAIRY AND SUGAR
American Sugar Beet Growers Association
International Association of Ice Cream Manufacturers (and Milk Industry
    Foundation)
National Independent Dairymen's Association
National Milk Producers Federation
Sugar Association of the Caribbean
Sugar Users Group

C. FEED, GRAINS AND OILS
Affiliated Rice Mills (Meyers and Associates)
American Bakers Association
Corn Refiners Association
National Association of Wheat Growers
National Corn Growers Association
Rice Millers' Association
U.S. Feed Grains Council

D. MEATS AND POULTRY
American Meat Institute
American Wool Growers Association

           Independent Cattlemen's Association (White and Associates)
           National Broiler Council
           National Cattlemen's Association
           National Pork Producers Council
           United Egg Producers (and United Egg Association; Scott, Harrison, and McLeod)

   E. SELECT CROPS
           Michigan Apple Committee
           Michigan Association of Cherry Producers
           United Fresh Fruit and Vegetable Association
           Wine Institute

V. COOPERATIVES
        National Cooperatives Business Association
        National Council of Farm Cooperatives
        National Rural Electric Cooperatives Association
        Sunkist Growers

VI. GENERAL FARM ORGANIZATIONS
        American Agri-Women
        American Farm Bureau Federation
        Communicating for Agriculture
        Midcontinent Farmers Association
        National Farmers Organization
        National Farmers Union
        National Grange
        U.S. Farmers Association
        Women involved in Farm Economics

VII. PUBLIC INSTITUTIONS
        National Association of State Departments of Agriculture
        National Association of State Universities and Land-Grant Colleges

VIII. RESEARCH INSTITUTIONS AND FOUNDATIONS
        American Enterprise Institute
        Center for National Policy
        Curry Foundation
        Farm Foundation
        Heritage Foundation
        National Center for Public Policy Research
        Resources for the Future

IIX. RURAL PROTEST GROUPS
        ACRES, USA
        American Agriculture Movement
        Family Farm Movement
        Groundswell
        Iowa Farm Unity Coalition
        Minnesota Farm Unity Coalition
        National Catholic Rural Life Conference
        National Organization for Raw Materials
        North American Farmers Alliance
        Prairiefire

IX. SELECT ISSUE ORGANIZATIONS

    A. CONSERVATION AND ENVIRONMENT
ACRES
American Farmland Trust
Conservation Foundation
Environmental Policy Institute
Institute for Alternative Agriculture
Soil Conservation Society
Worldwatch Institute

    B. CONSUMERS AND PUBLIC INTEREST
Bread for the World
Capitol Legal Foundation
Food Research and Action Center
Interfaith Action for Economic Justice
National Council of Senior Citizens
Public Voice

    C. FINANCE
Association of Family Farmers
Farm Credit Council
Independent Bankers Association of America

    D. LABOR
AFL-CIO
Farm Worker Justice Fund (Migrant Legal Action Program)
United Auto Workers

    E. RURAL
National Rural Housing Coalition
Rural America
Rural Coalition
Rural Governments Coalition

In addition, the following organized coalition/discussion groups were the specific subjects of inquiry with respondents from these Washington, D.C.-based forums:

Ag-Energy Users
Child Nutrition Forum
Coalition to Reduce Inflated Milk Prices
Coalition to Reduce Inflated Sugar Prices
Export Processing Industry Coalition
Farm Coalition
Food Group
National Farm Coalition
Pesticide Users Coalition

Appendix B
Interview Outline
for Interest Respondents

Request organization's policy interests

Probe for specific bills since 1980

Ask about priorities:
1. General issues
2. Bills/amendments/rulings

Ask about involvement in neglected issue areas (rotate specific examples for each issue area):
1. Price supports/farm income
2. International trade & markets (including world food)
3. Tax issues
4. Agricultural structure
5. Subsidized inputs
6. Conservation/food security
7. Nutrition/domestic hunger
8. Rural America/poverty
9. Health and safety

Probe relationship between policy positions and active lobbying

Ask about organizational resources and limitations on lobbying:
1. Personnel
2. Financial
3. Membership

Ask whom they work with:
1. Other organizations
2. Legislators
3. Administrators

Probe for degree of active involvement with these participants

Probe for coalition arrangements

Ask to assess impact of:
1. Coalitions
2. Organization
3. Agribusiness
4. Traditional farm interests
5. Conservationists
6. Consumers, urban interests, and others from past farm bills

Ask impressions of the politics of the 1985 farm bill

Request membership and staff data if not already available

Appendix C
Agricultural Policy
Decisions, 1980–1985

A.   USDA and International Food Trade Administrative Decisions

    1. Not to institute paid land diversion programs for feedgrains, 1980
    2. School lunch purchases of canned pork, 1980
    3. Wheat loan extension, 1980
    4. School lunch regulations, nutritional requirements, 1980
    5. Most-favored-nation status to China, 1980
    6. Lowered sugar import duty, 1980
    7. Processed potato purchases, 1980
    8. Barley reserve opened, 1980
    9. Regulation of DES-implanted cattle sales, 1980
  10. Changes in farmer-owned grain reserve, 1980
  11. Established upland cotton loan rate, 1980
  12. Marketing quota on extra-long-staple cotton, 1980
  13. Tobacco loan program, 1980
  14. Food stamp program regulations on job search, 1980
  15. California quarantine on fruit and vegetables for Medfly outbreaks, 1980
  16. Agreement on Mexican grain purchases, 1980
  17. Tariff on imported mushrooms, 1980
  18. Required exporter registration, 1980
  19. Restored meat imports from El Salvador, pesticide inspection, 1980
  20. Not to grant corn and sorghum deficiency payments, 1981
  21. Instituted 1982 wheat program, 1981
  22. Peanut support levels, 1981
  23. Brucellosis program changes, 1981
  24. Increased meat inspection fees, 1981
  25. Increased rates for Farmers Home Administration, 1981
  26. Regulation of processing on donated surplus foods for school lunch program, 1981
  27. Attached duties on Australian lamb, 1981

28. Creation of pool reserve under almond-marketing order, 1981
29. Established cereal import fund, 1981
30. Abolished Office of Environmental Quality, 1981
31. Sale of U.S. surplus butter to New Zealand, 1981
32. Voluntary reduction programs for feedgrains, 1982
33. Loan program changes, 1982
34. Procedure for deferring upland cotton loan differentials, 1982
35. Grain warehouse contracts and inspection fees, 1982
36. Peanut quota poundage levels, 1982
37. Terminated Puerto Rican tobacco purchase program, 1982
38. Interim rule on nonagenda tobacco, 1982
39. Revised sugar import fees, 1982
40. Set honey loan and purchase rates, 1982
41. Endorsed continued use of marketing orders, 1982
42. Raised fees on voluntary fruit and vegetable inspections, 1982
43. Revised grade standards for applesauce and apple juice, 1982
44. Cooperative agreement with Alaska for fruit and vegetable grading, 1982
45. Established mohair incentive payments, 1982
46. Terminated emergency feed program, 1982
47. State swine brucellosis regulations, 1982
48. Increased retail sales exemption for meat and poultry inspection, 1982
49. Eliminated special permits for returning faulty meat products to plants, 1982
50. Lowered grazing fees on public lands, 1982
51. Originated critical area soil and water conservation program, 1982
52. Extended fire ant quarantine, 1982
53. Expanded gypsy moth high risk zone, 1982
54. Allowed rural electric administration to sell tax write-offs, 1982
55. Allowed distribution of government owned cheese, 1982
56. Eliminated most wholesalers from Food Stamp program, 1982
57. Suspended Argentinian loan guarantees, Falkland crisis, 1982
58. U.S. and Jamaica barter agreement, 1982
59. Exchange of scientific and technical information with European countries, 1982
60. Continued involvement with International Science and Education Council, 1982
61. Eliminated 500 USDA publications, 1982
62. Acreage reduction on wheat program, 1982
63. Extended storage space agreements, 1982
64. Soybean loan extension, 1982
65. Excess tobacco poundage carryover, 1982
66. Regulations on misrepresentation of perishable commodities, 1982
67. Grade standards for kiwi fruit, 1982
68. Modified labeling of processed meat, 1982
69. Established test feeding programs for low-income elderly, 1982
70. Established blended credit program, 1982
71. Federal crop insurance extension of coverage area, 1982
72. Set tobacco marketing order levels, 1982
73. Flat-rate brucellosis indemnity payments for cattle, 1982
74. Designated Indianapolis as port for exporting livestock, 1982
75. Allowed use of autogenous biologies from other flocks and herds, 1983
76. Set same inspection standards for meat imports as domestic, 1983

77. Halted Czechoslovakian meat imports, PCB residues, 1983
78. Acreage conservation reserve reduction for feedgrains, 1983
79. Released farmer-owned reserves of corn, 1983
80. Allowed interstate rice movement for PIK program, 1983
81. Added popcorn to experimental crop insurance program, 1983
82. Revised flue-cured tobacco grade standards, 1983
83. Revised grade standards for tomatoes for processing, 1983
84. Exempted rendering plants from animal health regulations on waste fed to swine, 1983
85. Restricted horse imports from Italy, venereal disease, 1983
86. Increased 50-cent deduction for milk sold under dairy program, 1983
87. Set nonrecourse loan program for sugar, 1983
88. Developed youth volunteer program, Touch America Project, 1983
89. Implemented pesticide information retrieval system, 1983
90. Halted ground beef distribution of substandard cattle, 1983
91. Added rice to surplus food distribution, 1983
92. Signed five-year grain agreement with Soviet Union, 1983
93. Announced PL-480 commodities, 1983
94. Revised export credit guarantee program, 1983
95. Cooperative agreement with Brazil's national agricultural research organization, 1983
96. Lowered wheat program target price, 1984
97. Change rye standards, 1984
98. Adopted acreage conservation reserve for rice program, 1984
99. Changes in peanut program, 1984
100. Proposed federal sunflower standards, 1984
101. Eliminated "four leaf" tobacco program, 1984
102. Amended grade and quality standards to include all import tobacco, 1984
103. Changed grading for slaughtered lamb and sheep, 1984
104. Proposed speeding broiler and Cornish-hen inspection system, 1984
105. Modified meat and poultry inspection regulations, 1984
106. Regulation of curing solution in pork products by protein level, 1984
107. Quarantine for avian influenza, poultry watch instituted, 1984
108. Offered milk diversion program, 1984
109. Announced dairy cow slaughter report, 1984
110. Announced long-term conservation reserve, 1984
111. Added more weeds to import restriction list, 1984
112. Announced approved lender program for Farmers Home Administration, 1984
113. Began buying canned beef and pork for charities, 1984
114. Agreed to supply technical assistance to Guatemala, 1984
115. Banned meat imports from Chile, 1984
116. Adopted changes in Federal Seed Act for lawn and turf, 1984
117. Announced government-wide rural development policy, 1984
118. Eliminated moisture content as grade-determining factor, feedgrains, 1984
119. Proposed advertising changes in navel-orange marketing order, 1984
120. Temporarily amended federal milk marketing orders, 1984
121. Provided emergency feed assistance, 1984
122. Announced transitional program of additional farm credit, 1984
123. Formed Plant Gene Expression Center, speed genetic engeneering, 1984
124. Extended commodity program sign-up dates, 1985
125. Reduced grain industry record-keeping regulations, 1985

126. Revised grading standards for canned peaches, 1985
127. Closed U.S. border to Mexican citizens, 1985
128. Relaxed regulation of margarine sweeteners, 1985
129. Suspended import fee on raw sugar, 1985
130. Proposed regulations for importing livestock embryos, 1985
131. Regulation of pet-food manufacturers, greater flexibility, 1985
132. Allowed split manufacture of American produced biologics, 1985
133. Denied proposed amendments to milk-marketing orders, 1985
134. Announced boll weevil eradication program, Southwest, 1985
135. Declared gypsy moth emergency, Oregon, 1985
136. Approved farm debt set aside for Farmers Home Administration, 1985
137. Included interest write-down option in Farm Credit Relief program, 1985
138. Announced Special Producer Storage Loan Program, 1985
139. Continued scientific and technological exchanges in agriculture with Bulgaria, 1985
140. Improvements in Statistical and Economic Reporting Procedures, 1985

B. Legislative Action, Excluding Farm Bill Decisions

1. Extension of Emergency Farm Credit Adjustment Act, 1980
2. Raised Food Stamp authorization ceilings, 1980
3. Increased PL-480 aid, 1980
4. Farm Credit Act provision to authorize export credit, 1980
5. Farm Credit Act provision to rural commercial banks, 1980
6. Farm Credit Act provision to extend services to commercial fishers, 1980
7. Grain-weighing bill, 1980
8. Provided all-risk crop insurance, 1980
9. Designated undersecretary for Small Community and Rural Development, 1980
10. Delayed wheat referendum, 1981
11. Lowered dairy price support levels, 1981
12. Delayed increase in milk price supports, 1981
13. Required licensing of potato-processing firms, 1982
14. Mandated export assistance funding, 1982
15. Adjusted Thrifty Food Plan formula, 1982
16. No Net Cost Tobacco Program, 1982
17. Export Trading Company Act, antitrust protection, 1982
18. Reclamation Reform Act provision on free water, 1982
19. Funding to reinstitute monthly commodity reports, 1982
20. Extended Commodity Exchange Act, 1982
21. Surplus Agricultural Commodities Disposal Act, liquid fuel, 1982
22. Gave secretary of agriculture authority to sell National Forest lands, 1982
23. Established tax policy for Payment-In-Kind program, 1982
24. Caribbean Basin Economic Recovery Act, 1983
25. Amended tobacco price support levels, 1983
26. Standardized cotton programs, 1983
27. Extended International Coffee Agreement, 1983
28. Changed existing dairy program, 1983
29. Tobacco Adjustment Act of 1983
30. Changed Farmers Home Administration loan programs, 1984
31. Authorized marketing research and promotion for nuts, 1984
32. Potato research promotion, 1984

33. Provided legal protections for sellers of perishable commodities, 1984
34. Promoted commercial development of guayule, 1984
35. Provisions to aid farmers storing in bankrupt elevators, 1984
36. Agricultural Trade and Export Policy Commission Act, 1984
37. Cotton Statistics and Estimates Act, 1984
38. Supplemental appropriations for African famine relief, 1985
39. Authorized Commodity Credit Corporation payment of food-handling costs, 1985
40. Prohibited HHS ban on saccharine sales and distribution restrictions, 1985

Source: Policy Research Notes, July 1980–June 1985.

Notes

PREFACE

1. William P. Browne and Charles W. Wiggins, "Resolutions and Priorities: Lobbying by the General Farm Organizations," *Policy Studies Journal* 6 (Summer 1978): 493–498; William P. Browne and Charles W. Wiggins, "Interest Group Strength and Organizational Characteristics: The General Farm Organizations and the 1977 Farm Bill," in *The New Politics of Food*, eds., Don F. Hadwiger and William P. Browne (Lexington, Mass.: Lexington, 1978), pp. 109–121.

2. William P. Browne, "Farm Organizations and Agribusiness," in *Food Policy and Farm Programs*, eds., Don F. Hadwiger and Ross B. Talbot (New York: Academy of Political Science, 1982), pp. 198–211; Kenneth J. Meier and William P. Browne, "Interest Groups and Farm Structure," in *Farms in Transition*, eds., David Brewster, Wayne Rasmussen, and Garth Youngberg (Ames: Iowa State University Press, 1983), pp. 47–56; William P. Browne and Kenneth J. Meier, "Choosing Depletion? Soil Conservation and Agricultural Lobbying," in *Scarce Natural Resources: The Challenge to Public Policymaking*, eds., Susan Welch and Robert Miewald (Beverly Hills, Calif.: Sage, 1983), pp. 255–277.

3. William P. Browne, "Mobilizing and Activating Group Demands: The American Agriculture Movement," *Social Science Quarterly* 64 (March 1983): 19–34; William P. Browne and John Dinse, "The Emergence of the American Agriculture Movement, 1977–1979," *Great Plains Quarterly* 5 (Fall 1985): 221–235.

4. This 100 percent response rate would never have been possible without my affiliation with the Economic Research Service (ERS), U.S. Department of Agriculture. The reputation and prominence of ERS made it very difficult for policy participants to refuse me access. The nonthreatening nature of the questions elicited easy, comfortable, and open responses.

5. These decisions were sampled from periodic updates published in *Policy Research Notes* (Washington, D.C.: Economic Research Service, United States Department of Agriculture, 1980–1985).

CHAPTER 1. INTRODUCTION

1. David B. Truman, *The Governmental Process* (New York: Alfred A. Knopf, 1951); James Q. Wilson, *Political Organizations* (New York: Basic Books, 1973).

2. Andrew S. McFarland, *Common Cause: Lobbying in the Public Interest* (Chatham, N.J.: Chatham House, 1984).

3. Lester W. Milbrath, *The Washington Lobbyists* (Chicago: Rand McNally, 1963).

4. Charles W. Wiggins and William P. Browne, "Interest Groups and Public Policy within a State Legislative System," *Polity* 24 (Spring 1982): 548–558; L. Harmon Zeigler and Michael Baer, *Lobbying: Interaction and Influence within American State Legislatures* (Belmont, Calif.: Wadsworth, 1969).

5. Robert H. Salisbury, "Interest Representation: The Dominance of Institutions," *American Political Science Review* 78 (March 1984): 64–76; Keith H. Hamm, Charles W. Wiggins, and Charles G. Bell, "Interest Group Involvement, Conflict, and Success in State Legislatures," unpublished manuscript delivered at the Annual Meeting of the American Political Science Association (Chicago, 1983).

6. Graham Wootton, *Interest Groups: Policy and Politics in America* (Englewood Cliffs, N.J.: Prentice-Hall, 1985), p. 311.

7. Henry J. Pratt, *The Gray Lobby* (Chicago: University of Chicago Press, 1976).

8. Chinkook Lee, Gerald Schluter, William Edmondson, and Darryl Wills, "Measuring the Size of the U.S. Food and Fiber System," paper prepared for the Agricultural and Rural Economics Division, Economic Research Service, United States Department of Agriculture (Washington, D.C., undated).

9. These and other budget figures for this section, as updated, are from Geoffrey Becker and Jeffrey Zina, "The U.S. Department of Agriculture's FY86 Budget Request," paper prepared for the Food and Agricultural Section, Environment and Natural Resources Policy Division, Congressional Research Service, Library of Congress (Washington, D.C.: February 8, 1985).

10. Kenneth R. Farrell and C. Ford Runge, "Three Dimensions of U.S. Agricultural Policy," *Policy Research Notes* (December 1983): 13–19.

11. Lee et al., "Measuring the Size of the U.S. Food and Fiber System," p. 4.

12. Cecil W. Davison, "U.S. Agricultural Exports: Prospects for the 1980's," *Policy Research Notes* (August 1981): 13–23.

13. Jack L. Walker, "The Origins and Maintenance of Interest Groups in America," *American Political Science Review* 77 (June 1983): 390–406.

14. Don Paarlberg, *Farm and Food Policy: Issues of the 1980's* (Lincoln: University of Nebraska Press, 1980), pp. 59–64.

15. Jim Johnson, Kenneth Baum, and Richard Prescott, *Financial Characteristics of U.S. Farms, January 1985* (Washington, D.C.: U.S. Department of Agriculture, Economic Research Service, July 1985).

16. In the many conferences held by agricultural economists as a prelude to the 1985 Farm Bill, this was a near-consensus opinion. See, for example, Bruce L. Gardner, ed., *U.S. Agricultural Policy: The 1985 Farm Legislation* (Washington, D.C.: American Enterprise Institute, 1985).

17. Kay Lehman Schlozman and John T. Tierney, *Organized Interests and American Democracy* (New York: Harper and Row, 1986), p. 396.

18. James T. Bonnen, "Implications for Agricultural Policy," *American Journal of Agricultural Economics* 55 (August 1973): 391–398; for the effects of diffuse demands see Charles M. Hardin, *The Politics of Agriculture* (Glencoe, Ill.: Free Press, 1952).

19. Raymond A. Bauer, Ithiel de Sola Pool, and Lewis Anthony Dexter, *American Business and Public Policy* (New York: Atherton, 1963), pp. 323–331.

20. Weldon Barton, "Coalition-Building in the U.S. House of Representatives: Agricultural Legislation in 1973," in *Cases in Public Policy,* ed., James E. Anderson (New York: Praeger, 1976): 141–162; John G. Peters, "The 1977 Farm Bill: Coalitions in Congress," in Hadwiger and Browne, *New Politics of Food,* pp. 23–25; John G. Peters, "The 1981 Farm Bill," in Hadwiger and Talbot, *Food Policy and Farm Programs,* pp. 157–173.

CHAPTER 2. THE AGRICULTURAL LOBBY

1. Robert H. Salisbury, "An Exchange Theory of Interest Groups," *Midwest Journal of Political Science* 13 (February 1969): 1–32.

2. Walker, "Origins and Maintenance of Interest Groups in America."

3. This is based on initial membership increases of grassroots agricultural groups from the mid-nineteenth century through the first decade of the twentieth century. See Browne and Dinse, "Emergence of the American Agriculture Movement," p. 231.

4. Ibid., pp. 229–232. See also Solon Justus Buck, *The Granger Movement* (Cambridge, Mass.: Harvard University Press, 1913); Theodore Saloutos and John D. Hicks, *Agricultural Discontent in the Middle West, 1900–1939* (Madison: University of Wisconsin Press, 1951); Richard Hofstadter, *The Age of Reform* (New York: Alfred A. Knopf, 1953); Theodore Saloutos, *Farmer Movements in the South, 1865–1933* (Lincoln: University of Nebraska Press, 1960); Allen Weinstein, *Prelude to Populism: Origins of the Silver Issue, 1867–1878* (New Haven, Conn.: Yale University Press, 1970); Robert C. McMath, Jr., *Populist Vanguard: A History of the Southern Farmers' Alliance* (Chapel Hill: University of North Carolina Press, 1975).

5. John Mark Hansen, "Congressmen and Interest Groups: The Development of an Agricultural Policy Network in the 1920s," paper prepared for the Annual Meeting of the American Political Science Association (New Orleans, 1985), pp. 10–11; James H. Shideler, *Farm Crisis, 1919–1923* (Berkeley: University of California Press, 1957).

6. J. Clyde Marquis, "Farmers Can Have What They Want," *Country Gentleman* (November 1, 1919), p. 9.

7. Hansen, "Congressmen and Interest Groups," pp. 12–32.

8. L. Harmon Zeigler, *Interest Groups in American Society* (Englewood Cliffs, N.J.: Prentice-Hall, 1964), pp. 189–196.

9. O. M. Kile, *The Farm Bureau through Three Decades* (Baltimore: Waverly Press, 1948); Theodore J. Lowi, *The End of Liberalism,* 2d ed. (New York: Norton, 1979), pp. 68–75, 85–87.

10. This number is from a listing compiled from Arthur C. Close and Jody Curtis, eds., *Washington Representatives* (Washington, D.C.: Columbia Books, 1981–1985); Katherine Gruber, ed., *Encyclopedia of Associations,* Vol. 11, 20th ed. (Detroit: Gale Research Co., 1985); congressional hearings on farm, food, fiber, and foreign trade issues; *Washington Post* and *New York Times* files; and affiliate and subsidiary lists of national agricultural interests.

11. Harold D. Guither, *The Food Lobbyists* (Lexington, Mass.: Lexington Books, 1980).

12. Close and Curtis, *Washington Representatives,* 1985.

13. Wesley McCune, *Who's behind Our Farm Policy?* (New York: Frederick A. Praeger Publishers, 1956).

14. John P. Heinz, "The Political Impasse in Farm Support Legislation," *Yale Law Journal* 71 (April 1962): 954–970; Bonnen, "Implications for Agricultural Policy."

15. Mancur Olsen, Jr., *The Logic of Collective Action* (Cambridge, Mass.: Harvard University Press, 1965).

16. Salisbury, "Exchange Theory of Interest Groups," pp. 15–22.

17. The following works focus on agricultural interests and incentive theory: Olson, *Logic of Collective Action*, pp. 148–159; Salisbury, "Exchange Theory of Interest Groups"; J. Q. Wilson, *Political Organizations*, pp. 195–200; Terry M. Moe, *The Organization of Interests* (Chicago: University of Chicago Press, 1980), pp. 201–218; Browne, "Mobilizing and Activating Group Demands," pp. 19–34; Allan J. Cigler and John Mark Hansen, "Group Formation through Protest: The American Agriculture Movement," in *Interest Group Politics*, eds., Allan J. Cigler and Burdett A. Loomis (Washington, D.C.: Congressional Quarterly Press, 1983), pp. 84–109.

18. Louis Galambos, *Competition and Cooperation; The Emergence of a National Trade Organization* (Baltimore: Johns Hopkins University Press, 1966).

19. Truman, *Governmental Process*, p. 59.

20. Graham K. Wilson observes this partisan division well. See his *Special Interests and Policymaking: Agricultural Policies and Politics in Britain and the United States of America, 1957–70* (London: John Wiley, 1970).

21. William P. Browne, "Benefits and Membership: A Reappraisal of Interest Group Activity," *Western Political Quarterly* 29 (June 1976): 258–273; Moe, *Organization of Interests*, pp. 201–218.

22. For an excellent discussion of the meaning of interests, see Schlozman and Tierney, *Organized Interests and American Democracy*, pp. 16–23.

23. For a summary of the importance of latent interests, see L. Harmon Zeigler and G. Wayne Peak, *Interest Groups in American Society*, 2d ed. (Englewood Cliffs, N.J.: Prentice-Hall, 1972), pp. 3–4. Their distinction between interest groups and organized associations fails to account for the temporal and episodic nature of lobbying. For an explanation of far more relevance to the agricultural lobby, see Truman, *Governmental Process*, p. 33; and V. O. Key, *Public Opinion and American Democracy* (New York: Alfred A. Knopf, 1961), pp. 263–292. Key notes that while any "shared attitudes may become the basis of an interest group. . . . Some disturbance, though, may set off interactions among persons with those common attitudes" (p. 264).

24. Jeffrey M. Berry, *Lobbying for the People* (Princeton, N.J.: Princeton University Press, 1977), p. 7.

25. McCune, *Who's behind Our Farm Policy?*

26. The intent is not to be so descriptive as to make this book an encyclopedic effort. There are other sources to use for such detailed information. See Guither, *Food Lobbyists*; William P. Browne and Allan J. Cigler, eds., *Agricultural Interests in the United States* (Westport, Conn.: Greenwood Press, forthcoming).

27. After eliminating consultants from both lists, the 84 reputedly most active organizations accounted for 77 percent of my original list of 128 targeted respondent groups and firms (see Chapter 1). The falloff is explained by a drop in organizational involvement because of finances, relevance of issues, or credibility with decisionmakers. These factors are explained further in Chapters 4 through 7.

28. Milbrath, *Washington Lobbyists*, pp. 209-235.

29. Kenneth Towl, "Cargill, Inc.: Managing Corporate Public Policy in a Changing External Environment," paper prepared for the Harvard Business School (Boston: Intercollegiate Case Clearing House, 1977).

CHAPTER 3. THE POLICY PROCESS
AND INTEREST REPRESENTATION

1. Christiana M. Campbell, *The Farm Bureau and the New Deal* (Urbana: University

of Illinois Press, 1962); John A. Crompton, *The National Farmers Union* (Lincoln: University of Nebraska Press, 1965).

2. William P. Browne, "Policy and Interests: Instability and Change in a Classic Issue Subsystem," in *Interest Group Politics*, 2d ed., eds., Allan J. Cigler and Burdett Loomis (Washington, D.C.: Congressional Quarterly Press, 1986), pp. 183–201.

3. Richard F. Fenno, *Congressmen in Committee* (Boston: Little, Brown, 1973); John A. Ferejohn, *Pork Barrel Politics: Rivers and Harbor Legislation, 1947–1968* (Stanford, Calif.: Stanford University Press, 1974); Morris Fiorina, *Congress: Keystone to the Washington Establishment* (New Haven, Conn.: Yale University Press, 1977); Lawrence C. Dodd and Richard L. Schott, *Congress and the Administrative State* (New York: John Wiley, 1979), pp. 59–154.

4. Dodd and Schott, *Congress and the Administrative State*, pp. 276–323.

5. Paul H. Appleby, *Policy and Administration* (University: University of Alabama Press, 1949); Randall B. Ripley and Grace B. Franklin, *Bureaucracy and Public Policy*, 3d ed. (Homewood, Ill.: Dorsey Press, 1984), pp. 67–96.

6. The subgovernment or subsystems literature begins with the work of Ernest S. Griffith, *The Impasse of Democracy* (New York: Harrison-Hitton, 1939), and J. Leiper Freeman, *The Political Process: Executive Bureau-Legislative Committee Relations* (New York: Random House, 1955).

7. Barton, "Coalition-Building"; Peters, "1977 Farm Bill"; Peters, "1981 Farm Bill."

8. Theodore J. Lowi, *The End of Liberalism*, 2d ed. (New York: W. W. Norton, 1979).

9. Charles M. Hardin, "Agricultural Price Policy: The Political Role of Bureaucracy," *Policy Studies Journal* 6 (Summer 1978): 467–472; Clifford M. Hardin, "Congress Is the Problem," *Choices* 1 (January 1986): 6–10.

10. Neil L. Meyer and William T. Dishman, *Power Clusters: How Public Policy Originates* (Moscow, Idaho: College of Agriculture, University of Idaho, 1984). This work is directly based on Daniel M. Ogden, *How National Policy Is Made* (Washington, D.C.: Office of Power Marketing Coordination, U.S. Department of Education, 1983).

11. James T. Bonnen, "U.S. Agriculture, Instability and National Political Institutions: The Shift from Representative to Participatory Democracy," in *United States Agricultural Policies for 1985 and Beyond* (Tucson, Ariz.: Department of Agricultural Economics, University of Arizona, 1984), pp. 53–83.

12. The finding is confirmed by Schlozman and Tierney, *Organized Interests and American Democracy*, pp. 277–278.

13. Ferejohn, *Pork Barrel Politics*; Kenneth A. Shepsle, *The Giant Jigsaw Puzzle: Democratic Committee Assignments in the Modern House* (Chicago: University of Chicago Press, 1978).

14. Donald R. Matthews and James A. Stimson, *Yeas and Nays: Normal Decision-Making in the U.S. House of Representatives* (New York: John Wiley, 1975); John Kingdon, *Congressmen's Voting Decisions*, (New York: Harper and Row, 1981).

15. Schlozman and Tierney, *Organized Interests and American Democracy*, pp. 322–330.

16. Ibid., p. 272.

17. Bauer, de Sola Pool, and Dexter, *American Business and Public Policy*; Michael T. Hayes, *Lobbyists and Legislators* (New Brunswick, N.J.: Rutgers University Press, 1981); William P. Browne, "Variations in the Behavior and Style of State Lobbyists and Interest Groups," *Journal of Politics* 47 (March 1985): 450–468.

18. Milbrath, *The Washington Lobbyists*, pp. 209–294.

19. Harrison W. Fox, Jr., and Susan Webb Hammond, *Congressional Staffs: The Invisible Force in American Lawmaking* (New York: Free Press, 1977), pp. 12–32; Stephen J. Wayne, *The Legislative Presidency* (New York: Harper and Row, 1978), pp. 220–221;

David Nachmias and David H. Rosenbloom, *Bureaucratic Government USA* (New York: St. Martin's, 1980), p. 39; Louis W. Koenig, *The Chief Executive*, 4th ed. (New York: Harcourt Brace Jovanovich, 1981), p. 190; Robert H. Salisbury and Kenneth A. Shepsle, "Congressional Staff Turnover and the Ties-That-Bind," *American Political Science Review* 75 (June 1981): 381–425.

20. Robert H. Salisbury and Kenneth A. Shepsle, "U.S. Congressman as Enterprise," *Legislative Studies Quarterly* 6 (November 1981): 559–576.

21. The opinions of younger, perhaps newer, lobbyists are equally instructive. Although those from legislative staff and administrative backgrounds are highly content with staff interactions, lobbyists who come from backgrounds in state politics chafe a great deal. After years of working directly with state legislators, many express frustration and even anger with the lack of open doors in Congress.

22. Paul Gardner, "The Effectiveness of Farm Lobby Groups: A Congressional Survey for *Successful Farming*," unpublished manuscript (Ames: Department of Political Science, Iowa State University, 1979).

23. Browne and Dinse, "Emergence of the American Agricultural Movement," p. 232.

24. Allan J. Cigler, "From Protest Group to Interest Group: The Making of the American Agriculture Movement, Inc.," in Cigler and Loomis, *Interest Group Politics*, 2d ed., pp. 46–69.

25. These organizations best meet Truman's somewhat tautological observation that the group is defined in terms of its interest. That does not necessarily seem true of the two earlier-cited types. See *The Governmental Process*, pp. 33–39.

CHAPTER 4. AGRARIAN PROTEST
AND A GRASSROOTS LOBBY

1. Harold D. Guither, Bob F. Jones, Marshall A. Martin, and Robert G. F. Spitze, *U.S. Farmers' Views on Agricultural and Food Policy: A Seventeen-State Composite Report* (Ames, Iowa: North Central Region Extension, December 1984).

2. Michael S. Lewis-Beck, "Agrarian Political Behavior in the United States," *American Journal of Political Science* 21 (August 1977): 543–565.

3. Browne and Dinse, "Emergence of the American Agriculture Movement." See this article for a detailed account of the AAM's development. For an earlier account see Leo V. Mayer, "The Farm Strike," *Policy Research Notes* 6 (July 1978): 6–16.

4. Hugh J. Nolan, ed., *Pastoral Letters of the United States Catholic Bishops*, Vol. 3, 1962–1974 (Washington, D.C.: United States Catholic Conference, 1983), pp. 195–197, 404–405, 465; Hugh J. Nolan, ed., *Pastoral Letters of the United States Catholic Bishops*, Vol. 4, 1975–1983 (Washington, D.C.: United States Catholic Conference, 1984), pp. 47–53, 126; Committee on Social Development and World Peace, *The Family Farm* (Washington, D.C.: United States Catholic Conference, February 14, 1979).

5. Great Plains farmers seem to be suspicious of national politics anyway. A 1980 survey of Kansas farmers by Allan J. Cigler and John Mark Hansen showed that 82 percent of their respondents believed national government to be "run by a few big interests" who could be trusted only "some of the time," if at all. See "Group Formation through Protest."

6. Browne, "Mobilizing and Activating Group Demands," pp. 24–30.

7. These four paragraphs were derived from Browne and Dinse, "Emergence of the American Agriculture Movement," p. 222. There is a typographical error in that article on the number of attendees at the first strike meeting. It should read 40.

8. Aruna Nayer Michie and Craig Jagger found that 35 percent of Kansas Farm Bureau members identified themselves as AAM members. Another 27 percent labeled

themselves AAM sympathizers. Fifty-four percent of Kansas Farmers Union members joined AAM; another 21 percent sympathized with the movement. Of Kansas members of the National Farmers Organization, 66 percent joined AAM and another 24 percent expressed sympathy. See *Why Farmers Protest: Kansas Farmers, the Farm Problem, and the American Agriculture Movement* (Manhattan: Kansas State University, Agricultural Experiment Station, July 1980), pp. 38, 48, and 58.

9. Don F. Hadwiger, "Farmers in Politics," *Agricultural History* 50 (January 1976): 156–170. This tradition reappears on other farm issues when producers get no policy response. See Brian Wilson Coyer and Don S. Schwerin, "Bureaucratic Regulation and Farmer Protest in the Michigan PBB Contamination Case," *Rural Sociology* 46 (Winter 1981): 703–723.

10. Browne and Dinse, "Emergence of the American Agriculture Movement," p. 227.

11. Cigler, "From Protest Group to Interest Group"; Allan J. Cigler, "An Evolving Group in the Farm Crisis: Obstacles Facing the American Agriculture Movement," paper prepared for the Annual Meeting of the American Political Science Association (New Orleans, 1985).

12. A survey of AAM leader activists found that fewer than half of the respondents believed that a strike would succeed if organized. Moreover, one-half of those who thought a strike could work did not believe that farmers could get organized sufficiently to bring it off. Browne, "Mobilizing and Activating Group Demands." p. 24.

13. Mayer, "Farm Strike," p. 6.

14. Cigler, "From Protest Group to Interest Group," pp. 55–57.

15. Ibid., pp. 17–18.

16. Peter B. Clark and James Q. Wilson, "Incentive Systems: A Theory of Organizations," *Administrative Sciences Quarterly* 6 (September 1961): 129–166.

17. J. Q. Wilson, *Political Organizations*, p. 34.

18. Charles Tilly, *From Mobilization to Revolution* (Reading, Mass.: Addison-Wesley, 1978).

19. Browne and Dinse, "Emergence of the American Agricultural Movement," p. 232; Gilbert C. Fite, *American Farmers: First Majority, Last Minority* (Bloomington: Indiana University Press, 1981).

20. For a description of ideological beliefs, see Barbara A. Kohl, "Farm Movement Ideology in the Late Seventies: The American Agriculture Movement," unpublished paper (Columbus: Department of Agricultural Economics and Rural Sociology, Ohio State Universithy, 1979); see also Carl Boggs, "The New Populism and the Limits of Structural Reform," *Theory and Society* 12 (May 1983): 343–363.

21. Regional Catholic Bishops, *Strangers and Guests: Toward Community in the Heartland* (Sioux Falls, S. Dak.: Heartland Project, May 1, 1980).

22. Browne: "Variations in the Behavior and Style of State Lobbyists and Interest Groups," pp. 459–464.

23. Cigler and Hansen, "Group Formation through Protest," p. 106.

24. CBS News and *New York Times*, "The Farm Crisis," *CBS News Poll* (February 1986).

25. Doris A. Graber, *Mass Media and American Politics* (Washington, D.C.: Congressional Quarterly Press, 1980), p. 193.

CHAPTER 5. PRODUCERS IN WASHINGTON

1. Bauer, de Sola Pool, and Dexter, *American Business and Public Policy*, pp. 475–479.

2. David R. Mayhew suggests that these fears are well founded. See *Congress: The Electoral Connection* (New Haven, Conn.: Yale University Press, 1974), pp. 11–77.

3. David H. Harrington and Jerome M. Stam, *The Current Financial Condition of*

Farmers and Farm Lenders (Washington, D.C.: Economic Research Service, United States Department of Agriculture, March 1985).

4. Douglas E. Bowers, Wayne D. Rasmussen, and Gladys L. Baker, *History of Agricultural Price-Support and Adjustment Programs, 1933–1984* (Washington, D.C.: Economic Research Service, U.S. Department of Agriculture, December 1984.)

5. Thomas L. Gais, Mark A. Peterson, and Jack L. Walker, "Interest Groups, Iron Triangles, and Representative Institutions in American National Government," *British Journal of Political Science* 14 (March 1984): 161–185.

6. Wesley McCune, *The Farm Bloc* (Garden City, N.Y.: Doubleday, Doran, 1943), p. 188; Hardin, *Politics of Agriculture*.

7. Truman, *Governmental Process*, pp. 26–31.

8. James E. Anderson, "Agricultural Marketing Orders and the Process and Politics of Self-Regulation," *Policy Studies Review* 2 (August 1982): 97–111.

9. Bowers, Rasmussen, and Baker, *Agricultural Price-Support and Adjustment Programs*, p. 29.

10. Browne, "Farm Organizations and Agribusiness," p. 198.

11. This setting of priorities has been going on for at least ten years. See Browne and Wiggins, "Lobbying by General Farm Organizations."

12. Anthony Downs, *Inside Bureaucracy* (Boston: Little, Brown, 1967).

13. James E. Anderson, "Who Benefits from Farm Programs?" in *Food Policy and Farm Programs*, eds., Don F. Hadwiger and Ross B. Talbot (New York: Academy of Political Science, 1982): 144–156.

14. Jack Doyle, *Lines across the Land: Rural Electric Cooperatives* (Washington, D.C.: Environmental Policy Institute, 1979).

15. Jonathan Rauch, "Farmers' Discord over Government Role Produces a Farm Bill That Pleases Few," *National Journal* 17 (November 9, 1985): 2535–2536, 2538–2539.

16. Lyle P. Schertz and others, *Another Revolution in U.S. Farming?* (Washington, D.C.: Economics, Statistics, and Cooperatives Service, United States Department of Agriculture, December 1979), p. V.

17. Rauch, "Farm Bill Pleases Few," p. 2537.

CHAPTER 6. AGRIBUSINESS: LOBBYING AND ADJUSTMENT

1. A curious, indeed puzzling, error of analytical omission is the lack of attention that political scientists and policy-oriented agricultural economists have paid to agribusiness lobbies. The term *agribusiness* may well have been put into popular use by former (1971–1976) Secretary of Agriculture Earl Butz. As a phenomenon, however, American agribusiness is as old as the small towns that were founded in support of agricultural expansion. Farm journals show that farmers' complaints are long-standing. Historians such as Saloutos and Hicks (*Agricultural Discontent in the Middle West*) addressed farm organizations' dislike of businessmen, especially opinions expressed as part of populist ideology. There are also many historical references to extensive agribusiness involvement in such farm policy decisions as that on the Soil Bank of the 1950s. Have observers of agricultural policymaking been so directed toward mass membership interests that, after the robber barons gave way to some semblance of modern business, they failed to address adequately the question, Who helps move farm issues forward? After researching this project and reading the preliminary subsystems works of John Mark Hansen, I am inclined to believe so.

2. Larry G. Hamm, "Changes in Food Destination: Some Implications for Farm Policy

and Income," paper prepared for the Michigan Farm and Food Policy Conference (East Lansing: Michigan State University, 1983), p. 9.

3. Larry G. Hamm, "The Impact of Food Distributor Procurement Practices on Food System Structure and Coordination," working paper 58, NC Project 117, Studies of the Organization and Control of the U.S. Food System (Madison: University of Wisconsin, November 1981); Willard F. Mueller, "The Food Conglomerates," in Hadwiger and Talbot, *Food Policy and Farm Programs*, pp. 54–67; Larry G. Hamm and Gerald Grinnell, "Evolving Relationships between Food Manufacturers and Retailers: Implications for Food System Organization and Performance," *American Journal of Agricultural Economics* 65 (December 1983): 1065–1072.

4. Graham Wilson, "American Business and Politics," in Cigler and Loomis, *Interest Groups*, 2d ed., p. 221–235.

5. Dale C. Dahl and Jerome W. Hammond, *Market and Price Analysis: The Agricultural Industries* (New York: McGraw-Hill, 1977); Joel Solkoff, *The Politics of Food* (San Francisco: Sierra Club Books, 1985), pp. 81–116.

6. Lewis Anthony Dexter, *How Organizations Are Represented in Washington* (Indianapolis: Bobbs-Merrill, 1969), pp. 55–79.

7. This is in contrast to the broadly based, rather than single-industry, associations that James Q. Wilson found to be less involved in providing selective membership incentives (*Political Organizations*, p. 166).

8. For a CEO's view of government see Roger M. Blough, *The Washington Embrace of Business* (New York: Columbia University Press, 1975).

9. Dexter, *Organizations Represented in Washington*, p. 55.

10. Public relations activities, for this purpose, are explained well by Robert Engler, *The Politics of Oil: Private Power and Democratic Directions* (Chicago: University of Chicago Press, 1961), pp. 428–482.

11. "Cargill, Inc.," pp. 3–6; Solkoff, *Politics of Food*, p. 103.

12. A case study of tobacco can be seen in Robert H. Miles with Kim S. Cameron, *Coffin Nails and Corporate Strategies* (Englewood Cliffs, N.J.: Prentice-Hall, 1982), pp. 57–90.

13. The President's Task Force on International Private Enterprise, *Report to the President* (Washington, D.C.: December 1984). The Agency for International Development supported this report, a reminder of a long-standing business strategy of using government to advise government about corporate needs. See Mark V. Nadel, *Corporations and Political Accountability* (Lexington, Mass.: D. C. Heath, 1976), pp. 53–56.

14. In this regard, Cargill seems to be a typical business lobby. See Nadel, *Corporations and Political Accountability*, pp. 65–71.

15. A survey was sent to agricultural staff employees from congressional agricultural committees, the agricultural appropriations subcommittees, and individual offices of House and Senate agricultural committee members. The intent was to determine which organizations were recognizable for their roles in agriculture policy. It was a useful but not a central feature of this study. Of approximately fifty questionnaires sent out by Congressman Bill Schuette's House staff, twenty-three were returned either to them or directly to me.

16. Other indicators of staff knowledge were built into the legislative staff questionnaire, such as a few nonlobbies that had names much like those of active lobbies. These groups were frequently mistaken to be both actively involved and influential. Only one respondent noted specifically that that group was not allowed to lobby. Only one other respondent correctly pointed out that the name of a grain firm was incorrectly spelled and a trade association misnamed, both purposely. All of these instances added to my own and

the respondents' feelings that not a great deal of information was known about many of the organizations that come calling on Capitol Hill.

17. It seems unlikely that these are all the instances in which such departures were forthcoming from either the agribusiness or farm lobbies. The figures, however, do indicate some magnitude of the operational differences between agribusiness and producer organizations even if the more circumspect policy approach of agribusiness disguised many changing positions, especially within the politics of trade associations.

18. Salisbury, "An Exchange Theory of Interest Groups."

19. Barton, "Coalition Building"; Craig L. Infanger, William C. Bailey, and David R. Dyer, "Agricultural Policy in Austerity: The Making of the 1981 Farm Bill," *American Journal of Agricultural Economics* 65 (February 1983): 1–9.

20. Robert H. Salisbury, "Interest Groups," in *Handbook of Political Science*, Vol. 4, eds., Fred Greenstein and Nelson Polsby (Reading, Mass.: Addison-Wesley, 1975), pp. 171–228.

CHAPTER 7. THE OTHER LOBBY: PRIVATE VIEWS WITH A PUBLIC INTEREST

1. Guither, *Food Lobbyists*, p. 87.

2. Don F. Hadwiger, *The Politics of Agricultural Research* (Lincoln: University of Nebraska Press, 1982), pp. 150–168.

3. This situation is most reminiscent of the problem identified by Edward C. Banfield and James Q. Wilson as they looked at the two prevailing concepts of the urban public interest in *City Politics* (New York: Random House, 1963), p. 46.

4. Jim Hightower with Susan DeMarco, *Hard Tomatoes, Hard Times* (Cambridge, Mass.: Schenkman Publishing, 1973).

5. Andrew S. McFarland, *Public Interest Lobbies: Decision Making on Energy* (Washington, D.C.: American Enterprise Institute, 1976), pp. 2–5.

6. John Strohm, "Fifty Years of NWF History: Rocky Years, Then Explosive Growth," *National Wildlife* (April–May 1986), p. 53.

7. Founding dates are listed for these organizations in Gruber, *Encyclopedia of Associations*, Vols. 1 and 2.

8. John W. Kingdon, *Agendas, Alternatives, and Public Policies* (Boston: Little, Brown, 1984), pp. 173–204.

9. Greg Calvert, "A Left-Wing Alternative," in *The Case for Participatory Democracy*, eds., C. George Benello and Dimitrios Roussopoulos (New York: Viking Press, 1971), p. 377.

10. Ibid., p. 378.

11. Berry, *Lobbying for the People*, pp. 45–78.

12. Guither, *Food Lobbyists*, p. 88.

13. Stuart Langton, "The Future of the Environmental Movement," in *Environmental Leadership*, ed., Stuart Langton (Lexington, Mass.: Lexington, 1984), p. 1.

14. Theodore Jacquency, "Washington Pressures: Public Interest Groups Challenge Government, Industry," *National Journal* 6 (February 23, 1974), p. 267.

15. Nick Kotz, *Let Them Eat Promises: The Politics of Hunger in America* (Garden City, N.Y.: Doubleday, 1971), pp. 46–55.

16. Jacquency, "Washington Pressures," pp. 267–268.

17. Jeffrey M. Berry, *Feeding Hungry People: Rulemaking in the Food Stamp Program* (New Brunswick, N.J.: Rutgers University Press, 1984), p. 91.

18. Michael Pertschuk, *Giant Killers* (New York: Norton, 1986).

19. The Cornucopia Project, *Empty Breadbasket? The Coming Challenge to America's Food Supply and What We Can Do about It* (Emmaus, Pa.: Rodale Press, 1981).

20. Hadwiger, *Politics of Agricultural Research*, pp. 166–167.

21. Jack Doyle, *Altered Harvest: Agriculture, Genetics, and the Fate of the World's Food Supply* (New York: Viking, 1985).

22. Previous research indicated that USDA officials were especially uncooperative with public interest groups. See Berry, *Lobbying for the People*, p. 216.

23. Tony Smith, "Science and Ideology in Agriculture," paper presented at the Third Annual Conference on Religious Ethics and Technological Change, Iowa State University, Ames (February 1986).

24. United States Department of Agriculture, *1982 Census of Agriculture: Preliminary Report* (Washington, D.C.: USDA, May 1984).

25. For a more detailed account, see Christopher J. Bosso, "Transforming Adversaries into Collaborators: Interest Groups and the Regulation of Chemical Pesticides," paper presented at the Annual Meeting of the American Political Science Association (Washington, D.C.: 1986).

26. Oran R. Young, *Resource Regimes: Natural Resources and Social Institutions* (Berkeley: University of California Press, 1982), pp. 45–67.

27. Peters, "The 1977 Farm Bill," 24–25, 31–32.

28. Berry, *Lobbying for the People*, pp. 203–204.

CHAPTER 8. CONSULTANTS AND EXPERTS

1. Milbrath, *Washington Lobbyists*, pp. 159–161.

2. McCune, *Who's behind Our Farm Policy?*

3. "The Best Known Lobbyists Are Often Seen as Most Overrated," *National Journal* 17 (September 14, 1985): 2034–2035.

4. Ibid.

5. Guither, *Food Lobbyists*, pp. 81–82.

6. Ann Cooper, "Lobbying in the '80's: High Tech Takes Hold," *National Journal* 17 (September 14, 1985): 2030–2032.

7. In this sense, these consultants are much like the bureaucratic policy analysts described by Arnold J. Meltsner in *Policy Analysts in the Bureaucracy* (Berkeley: University of California Press, 1976), pp. 24–27.

8. The firm of Abel, Daft, and Earley is the organizational successor to Schnittker and Associates after John Schnittker's semiretirment to part-time practice.

9. Meltsner, *Policy Analysts in the Bureaucracy*, pp. 21–22.

10. "The Billion-Dollar Farm Co-Ops Nobody Knows," *Business Week* (February 7, 1977): 54–60.

11. Guither, *Food Lobbyists*, pp. 60–67.

12. Kay Lehman Schlozman and John T. Tierney, "More of the Same: Washington Pressure Activity in a Decade of Change," *Journal of Politics* 45 (May 1983): 351–377.

13. Graham Wilson raises the point that in general the political activity of business has been deceptively low in relationship to its prominence. This seems another case in point. See "American Business and Politics" and his *Business and Politics: A Comparative Introduction* (Chatham, N.J.: Chatham House, 1985).

14. Langton, "Networking and the Environmental Movement," pp. 132–133.

15. Hadwiger, *Politics of Agricultural Research*, pp. 32–67.

16. Bruce L. R. Smith, *The RAND Corporation* (Cambridge, Mass.: Harvard University Press, 1986), p. 31.

CHAPTER 9. COALITION POLITICS:
DILEMMAS AND CHOICES

1. J. Q. Wilson, *Political Organizations*, pp. 261–280.

2. William P. Browne, "Organizational Maintenance: The Internal Operation of Interest Groups," *Public Administration Review* 37 (January–February 1977): 54–55.

3. Bill Keller, "Coalitions and Associations Transform Strategy: Methods of Lobbying in Washington," *National Journal* 14 (January 23, 1982): 119–123.

4. Schlozman and Tierney, *Organized Interest and American Democracy*, p. 155.

5. Robert H. Salisbury, John P. Heinz, Edward O. Lauman, and Robert L. Nelson, "Who Works with Whom? Patterns of Interest Group Alliance and Opposition," paper presented at the Annual Meeting of the American Political Science Association (Washington, D.C., 1986).

6. J. Q. Wilson, *Political Organizations*, p. 267.

7. William H. Riker, *The Theory of Political Coalitions* (New Haven, Conn.: Yale University Press, 1962), pp. 255–256.

8. Salisbury et al., "Who Works with Whom?" p. 8.

9. Bowers, Rasmussen, and Baker, *History of Agricultural Price-Support and Adjustment Programs*, p. 21.

10. Garth Youngberg. "The National Farm Coalition and the Politics of Food," paper prepared for the Fourth Annual Hendricks Public Policy Symposium (Lincoln, Nebr., 1979).

11. James T. Bonnen, "Agriculture in the Information Age," paper prepared for the Agricultural Institute of Canada (Saskatoon, July 7, 1986), especially pp. 7–19.

12. Anne N. Costain and W. Douglas Costain, "Interest Groups as Aggregators in the Legislative Process," *Polity* 14 (Winter 1981): 249–272.

13. Loomis, "Coalitions of Interests: Building Bridges in the Balkanized State," in Cigler and Loomis, *Interest Group Politics*, 2d ed., pp. 258–274.

14. See also Schlozman and Tierney, *Organized Interests and American Democracy*, for the difficulties of meeting this ideal.

15. J. Q. Wilson, *Political Organizations*, p. 278.

16. Gilbert C. Fite, "The Changing Political Role of the Farmer," *Current History* 31 (August 1956): 84–90.

17. McCune, *Who's behind Our Farm Policy?* pp. 3–13.

18. These responses help explain why representatives of similar kinds of groups, such as commodity organizations, report that they most often work together. See Salisbury et al., "Who Works with Whom?" pp. 19–21.

19. For a detailed commentary on these conditions, their causes and effects, see *Increasing Understanding of Public Policies and Problems—1986* (Oak Brook, Ill.: Farm Foundation, 1987).

20. Salisbury, "An Exchange Theory of Interest Groups"; J. Q. Wilson, *Political Organizations*, pp. 30–35.

21. Loomis, "Coalitions of Interests."

CHAPTER 10. ISSUES, INTERESTS,
AND PUBLIC OFFICIALS

1. While these criticism owe mostly to the farm protest groups, especially the AAM,

they are brought up in academic analyses of what makes the AFBF work. Olson, *Logic of Collective Action*, pp. 153–139; Salisbury, "An Exchange Theory of Interest Groups."

2. Robert H. Salisbury, "Interest Groups," 171–128.

3. See also Berry, *Lobbying for the People*, pp. 59–76.

4. Hadwiger, *Politics of Agricultural Research*, pp. 90–168.

5. Wiggins and Browne, "Interest Groups and Public Policy"; Charles W. Wiggins, "Interest Group Involvement and Success within a State Legislative System," in *Public Opinion and Public Policy*, 3d ed., ed. Norman P. Luttbeg (Itasca, Ill.: Peacock, 1981): 216–239; Charles W. Wiggins, Keith E. Hamm, and Charles G. Bell, "Interest Groups and Other Influence Agents in the State Legislative Process: A Comparative Assessment," paper prepared for Annual Meeting of the American Political Science Association (Washington, D.C: 1984).

6. Kenneth J. Meier and J. R. Van Loherizen, "Interest Groups in the Appropriations Process: 'The Wasted Profession' Revisited," *Social Science Quarterly* 59 (December 1978): 482–495.

7. Hadwiger, *Politics of Agricultural Research*; Don F. Hadwiger and William P. Browne, eds., *Public Policy and Agricultural Technology: Adversity despite Achievement* (London: Macmillan, 1987).

8. These four issues were selected randomly from a list of twenty major policy conflicts in agriculture between 1983 and 1985. The list was prepared by ERS-USDA staff. Time allowed for only four issues to be selected.

9. Don Paarlberg, *American Farm Policy* (New York: Wiley, 1964), p. 116.

10. These points are the same as reported in earlier research on the general farm organizations. See Browne and Wiggins, "Interest Group Strength and Organizational Characteristics," p. 114.

11. Salisbury, "An Exchange Theory of Interest Groups."

12. George Brandsberg, *The Two Sides in NFO's Battle* (Ames: Iowa State University Press, 1964); Don F. Hadwiger and Ross B. Talbot, *Pressures and Protests: The Kennedy Farm Program and the Wheat Referendum of 1963* (San Francisco: Chandler, 1965); Harold G. Halcrow, *Food Policy for America* (New York: McGraw-Hill, 1978).

13. Stephen B. Sarason and Vera H. Sarason, *Political Party Patterns in Michigan* (Detroit: Wayne State University Press, 1957); J. David Greenstone, *Labor in American Politics* (New York: Alfred A. Knopf, 1969).

14. The phrase but not the conclusion belongs to Jeffrey M. Berry, *The Interest Group Society* (Boston: Little, Brown, 1984), p. 212.

15. Kenneth J. Meier, *Regulation: Politics, Bureaucracy, and Economics* (New York: St. Martin's Press, 1985), pp. 119–138.

16. See L. Harmon Zeigler, "The Effects of Lobbying: A Comparative Assessment," in Luttbeg, *Public Opinion and Public Policy*, pp. 203–205, for an explanation of these contrasting views.

17. Bauer, de Sola Pool, and Dexter, *American Business and Public Policy*, pp. 258–276.

18. Hayes, *Lobbyists and Legislators*, pp. 19–39; Wiggins and Browne, "Interest Groups and Public Policy within a State Legislative Setting," pp. 254–256.

19. Wiggins, Hamm, and Bell, "Interest Groups and Other Influence Agents," p. 22.

20. Bauer, de Sola Pool, and Dexter, *American Business and Public Policy*, pp. 127–153.

21. As explained in the Preface, the source was *Policy Research Notes*. Unfortunately, no source compiled an exhaustive list of administrative rulings and legislation posited and rejected, so these defeated items could not be included in any systematic way. There is some concern that this source was not as thorough in 1980 and 1981 as it was in later years. The decisions included were still adequate for the purpose, however.

22. In both instances, distinct provisions (as listed in *Policy Research Notes*) were counted as decisions, rather than considered a single item. On the other hand, administrative rulings that were made in more than a single year were lumped as one item if no policy changes were made. Only the first year is listed in Appendix C.

23. While few respondents did not have time to talk about specific issues, or were reluctant to do so, multiple interviews proved successful in collecting these data for 102 organizations including almost every organization on the list of most active national agricultural lobbies (Table 2.1). There is certainly no claim that this information reflects every issue involving each respondent organization.

24. This research design was patterned after the Wiggins's studies but is less revealing for two reasons. First, Washington lobbyists do not have to register and state positions on each bill of concern to them. Thus, follow-through was far more difficult. Second, because administrative rulings were included and because of the size of the bureaucracy, access to those who may initially have promoted the decision was much harder to arrange. In attempts to identify which policymakers made an initial proposal, the overwhelming response from most government respondents who were not directly involved was, "Who knows?" The size and decentralization of government made cross-checking of information between lobbyists and legislators virtually impossible. This is unfortunate since so many researchers report inconsistent responses between these sets of informants. See Zeigler, "Effects of Lobbying," p. 204.

25. Olson, *Logic of Collective Action*.

26. Here, as in other classifications within this chapter, there may be minor differences.

27. J. B. Penn, "An Appraisal of Policy Education Efforts for 1985 Agricultural legislation," in *Increasing Understanding of Public Policies and Problems—1986*.

28. Since political scientists have become accustomed to explaining the policy process as largely noncontroversial, this finding is hardly surprising. For a good review, see Dodd and Schott, *Congress and the Administrative State*, p. 276–323.

29. Robert H. Salisbury, "The Analysis of Public Policy: A Search for Theories and Rules," in *Political Science and Public Policy*, ed., Austin Ranney (Chicago: Markham, 1968): 151–175; Robert H. Salisbury and John P. Heinz, "A Theory of Policy Analysis and Some Preliminary Applications," in *Policy Analysis in Political Science*, ed., Ira Sharkansky (Chicago: Markham, 1970): 39–60.

30. Clearly commodity support programs have become redistributive public policy in the 1980s rather than distributive programs in which each farmer always gets the same or more. In the first place, policymakers worked with farm support package caps, or ceilings, as a constraint. Then it was decided who got what. Second, decisions in Congress and the USDA are intended to and do cut, trim, and impose costs on producer benefits so that different farmers constantly win and lose. A reserve or storage loan program may or may not last. Finally, commodity groups have behaved as if these policies were redistributive, each one always trying to get a better deal than other commodity organizations got. In the 1980s, these groups have not agreed on a mutual price support package or on other procedures; instead they elect to go with one or two coalition partners. One must look beyond the escalating total costs of farm programs in making this decision not to consider them distributive policy. Another factor to consider is the ongoing redistribution of benefits by region as more productive farmers, regionally distributed, get more from farm programs and less productive peers lose profits.

31. Heinz Eulau did not contend, nor would executives who hire them, that lobbyists are unimportant. See "Lobbyists: The Wasted Profession," *Public Opinion Quarterly* 28 (Spring 1964), p. 35, for a discussion of their role.

CHAPTER 11. PRIVATE INTERESTS
AND THE FOOD SECURITY ACT OF 1985

1. For more on the politics of ideas see Martha Derthick and Paul J. Quirk, *The Politics of Deregulation* (Washington, D.C.: Brookings, 1985).

2. James T. Bonnen, "Observations on the Changing Nature of National Agricultural Policy Decision Processes, 1946–76," in *Farmers, Bureaucrats, and Middlemen: Perspectives on American Agriculture*, ed., Trudy Huskamp Peterson (Washington, D.C.: Howard University Press, 1980): 309–328; Douglas E. Bowers, "Participants in the 1985 Farm Bill Legislation," unpublished manuscript (Washington, D.C.: Agricultural and Rural History Branch, Agricultural and Rural Economics Division, Economic Research Service, U.S. Department of Agriculture, August 11, 1986); Christopher K. Leman and Robert L. Paarlberg, "The Continued Political Power of Agricultural Interests," unpublished manuscript (Seattle: Graduate School of Public Affairs, University of Washington, 1985).

3. Bowers, "Participants in the 1985 Farm Bill Legislation." p. 4.

4. Bowers, Rasmussen, and Baker, *Agricultural Price-Support and Adjustment Programs*, pp. 3–5.

5. Lowi, *End of Liberalism*, pp. 85–87.

6. Heinz, "Political Impasse in Farm Support Legislation."

7. Peters, "1981 Farm Bill," p. 169.

8. Barton, "Coalition Building in U.S. House of Representatives: Agricultural Legislation."

9. Peters, "1977 Farm Bill."

10. Young Executives' Committee, "USDA and Food Policy Decisionmaking," unpublished mimeo (Washington, D.C.: U.S. Department of Agriculture, 1976), p. 6.

11. R. G. F. Spitze, "The Agriculture and Food Act of 1981: Continued Policy Evolution," *North Central Journal of Agricultural Economics* 5 (July 1983): 69.

12. Ross B. Talbot, "The Role of World Food Organizations," in *World Food Policies: Toward Agricultural Interdependence*, eds., William P. Browne and Don F. Hadwiger (Boulder, Colo.: Lynne Rienner Publishers, 1986), pp. 180–182.

13. John E. Lee, Jr., and Economic Indicators Branch Staff, "Financial Prospects for the Farm Sector," paper prepared for the Annual Agricultural Outlook Conference, U.S. Department of Agriculture (Washington, D.C., December 3, 1985).

14. Bruce Stokes, "A Divided Farm Lobby," *National Journal* (March 23, 1985): 632–638; Kent A. Price, ed., *The Dilemmas of Choice* (Washington, D.C.: Resources for the Future, 1985).

15. Lynn Daft, "The 1985 Farm Bill," *Choices* 1 (Premier Edition, 1986), p. 40.

16. McCune was one of the first to point this out. See *Who's behind Our Farm Policy?*

17. Willard W. Cochrane, "A New Sheet of Music," *Choices* 1 (Premier Edition 1986): 11–15.

18. McCune, *Who's behind Our Farm Policy?* pp. 149–155.

19. Two contrasting examples of book length reports include: Gardner, *U.S. Agricultural Policy: The 1985 Farm Legislation:* William A. Galston, *A Tough Row to Hoe: The 1985 Farm Bill and Beyond* (Washington D.C.: Hamilton Press/Roosevelt Center for American Policy Studies, 1985).

20. Barbara C. Stucker and Keith J. Collins, *The Food Security Act of 1985: Major Provisions Affecting Commodities* (Washington, D.C.: National Economics Division, Economic Research Service, U.S. Department of Agriculture, January 1986).

21. Lewrene K. Glaser, *Provisions of the Food Security Act of 1985* (National Economics Division, Economic Research Service, U.S. Department of Agriculture, 1985).

22. Bonnen, "Implications for Agricultural Policy," p. 398.

23. Keith Collins and Larry Saluthe, "Implications for Grains under the 1985 Farm Bill: A Comparison with the 1981 Act Authorities," paper prepared for acceptance at the Annual Meeting of the American Agricultural Economics Association (1986), p. 7. Their simulation revealed only "slightly less planted acreage and production under the 1985 Act."

24. American Farmland Trust, *Soil Conservation in America: What Do We Have to Lose?* (Washington, D.C.: American Farmland Trust, 1984).

25. Stucker and Collins, *Food Security Act of 1985*, p. 21. See remarks throughout the 1986 *Congressional Record* attributed to Senators William Armstrong (R-Colo.) and Richard Lugar (R-Ind.) and Representatives Tom Daschle (D-S.D.), Dan Glickman (D-Kans.), Robert Kasten (R-Wis.), and Howard Wolpe (D-Mich.).

26. Sandra J. Batie, Leonard A. Shabman, and Randall A. Kramer, "U.S. Agriculture and Natural Resources Policy," in Price, *The Dilemmas of Choice*, pp. 127–146.

27. For the blow-by-blow major commodity program battles see Bowers, *Participants in the 1985 Farm Bill Legislation*, pp. 8–11.

28. Daft, "1985 Farm Bill," p. 41.

29. This corroborates much of what William T. Gormley, Jr., concluded in "Regulatory Issue Networks in a Federal System," *Polity* 18 (Summer 1986): 595–620.

CHAPTER 12. ON INFLUENCING PUBLIC POLICY

1. Paarlberg, *Farm and Food Policy*, pp. 14–19.

2. Bonnen, "U.S. Agriculture, Instability, and National Political Institutions," pp. 54–55.

3. Ibid., p. 55.

4. Bowers, Rasmussen, and Baker, *Agricultural Price-Support and Adjustment Programs*.

5. For an excellent analysis of agriculture's special characteristics, see Bruce L. Gardner, *The Governing of Agriculture* (Lawrence: University Press of Kansas, 1981), pp. 85–101.

6. Alden C. Manchester, *The Public Role in the Dairy Economy: Why and How Governments Intervene in the Milk Business* (Boulder, Colo.: Westview Press, 1983), pp. 3–8.

7. Hardin, "Agricultural Price Policy," p. 469–471.

8. Clarence D. Palmby, *Made in Washington* (Danville, Ill.: Interstate, 1985), pp. 1–7.

9. Willard Cochrane and Mary E. Ryan, *American Farm Policy, 1948–1973* (Minneapolis: University of Minnesota Press, 1976), p. 114; Paarlberg, *Farm and Food Policy*, pp. 42–51.

10. Bonnen, "U.S. Agriculture, Instability, and National Political Institutions," pp. 81–82.

11. Paarlberg, "Farm Policy Agenda."

12. Frederick W. Taylor, *The Principles of Scientific Management* (New York: Harper and Brothers, 1911).

13. Raymond E. Callahan, *Education and the Cult of Efficiency* (Chicago: University of Chicago Press, 1962); Martin J. Schiesl, *The Politics of Efficiency: Municipal Administration and Reform in America, 1880–1920* (Los Angeles: University of California Press, 1977).

14. J. Q. Wilson, *Political Organizations*; Tilly, *From Mobilization to Revolution*.

15. Bauer, de Sola Pool, and Dexter, *American Business and Public Policy*, pp. 324–325.

16. J. Q. Wilson, *Political Organizations*, pp. 327–346.

17. This seems preferable to considering any nonparty organization an interest group as long it was "distinguishable by class of behavior, or activity" in attempting to influence public policy. Not all organizations behave alike. In addition, policymaker respondents do distinguish between interest groups and other forms of private interests. Wootton, *Interest Group Policy and Politics in America*, p. 23.

18. Olson, *Logic of Collective Action*, p. 126.

19. Salisbury, "An Exchange Theory of Interest Groups," pp. 38–41.

20. William A. Gamson, *The Strategy of Social Protest* (Homewood, Ill.: Dorsey, 1973), pp. 68–71.

21. Lawrence B. Mohr, "The Concept of Organizational Goal," *American Political Science Review* 77 (June 1973): 470–481.

22. Hugh Heclo, "Issue Networks and the Executive Establishment," pp. 113–115.

23. Michael P. Hayes, "The New Group Universe," in Cigler and Loomis, *Interest Group Politics*, 2d. ed., pp. 133–145.

24. For more elaboration on this point see William P. Browne, "Policy and Interests: Instability and Change in a Classic Issue Subsystem," in Cigler and Loomis, *Interest Group Politics*, 2d ed., pp. 183–201.

25. Schlozman and Tierney, "More of the Same."

26. Salisbury, "Interest Representation"; Kay Lehman Schlozman, "What Accents the Heavenly Chorus? Political Equality and the American Pressure System," *Journal of Politics* 46 (November 1984): 1006–1032.

27. Walker, "Origins and Maintenance of Interest Groups in America."

28. G. K. Wilson, *Special Interests and Policymaking*.

29. Marshall A. Martin, Harold D. Guither, and Robert G. F. Spitz, "Most Farmers Got Much of What They Wanted," *Choices* 1 (Third Quarter 1986): 34–35.

30. Earl Latham, *The Group Basis of Politics: A Study of Basing-Point Legislation* (Ithaca, N.Y.: Cornell University Press, 1952).

31. E. E. Schattschneider, *The Semi-Sovereign People* (New York: Holt, Rinehart, and Winston, 1960), p. 3.

32. However, the same fragmentation of interests seems to be taking place now in Europe and raising similar concerns over governance. See Suzanne P. Berger, ed., *Organizing Interests in Western Europe* (London: Cambridge University Press, 1981).

33. Salisbury, "Interest Groups," p. 180.

34. Hayes, *Lobbyists and Legislators*, p. 25.

Index